Topics in Non-Gaussian Signal Processing

Edward J. Wegman Stuart C. Schwartz
John B. Thomas
Editors

Topics in Non-Gaussian Signal Processing

With 83 Illustrations

Springer-Verlag
New York Berlin Heidelberg
London Paris Tokyo

A Dowden & Culver Book

Edward J. Wegman
Center for Computational Statistics
and Probability
George Mason University
Fairfax, VA 22030
USA

Stuart C. Schwartz
John B. Thomas
Department of Electrical Engineering
Princeton University
Princeton, NJ 08540
USA

Library of Congress Cataloging-in-Publication Data
Topics in non-Gaussian signal processing / edited by Edward J. Wegman,
Stuart C. Schwartz, John B. Thomas.
p. cm.
1. Signal processing—Digital techniques. I. Wegman, Edward J.,
II. Schwartz, Stuart C. (Stuart Carl), III. Thomas, John Bowman.
TK5102.5.T62 1988
621.38′043—dc19 87-33699

Printed on acid-free paper

Camera-ready copy prepared by the editors.
Printed and bound by R.R. Donnelley and Sons, Harrisonburg, Virginia.
Printed in the United States of America.

9 8 7 6 5 4 3 2 1

ISBN 0-387-96927-6 Springer-Verlag New York Berlin Heidelberg
ISBN 3-540-96927-6 Springer-Verlag Berlin Heidelberg New York

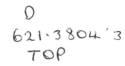

PREFACE

Non-Gaussian Signal Processing is a child of a technological push. It is evident that we are moving from an era of simple signal processing with relatively primitive electronic circuits to one in which digital processing systems, in a combined hardware-software configuration, are quite capable of implementing advanced mathematical and statistical procedures. Moreover, as these processing techniques become more sophisticated and powerful, the sharper resolution of the resulting system brings into question the classic distributional assumptions of Gaussianity for both noise and signal processes. This in turn opens the door to a fundamental reexamination of structure and inference methods for non-Gaussian stochastic processes together with the application of such processes as models in the context of filtering, estimation, detection and signal extraction. Based on the premise that such a fundamental reexamination was timely, in 1981 the Office of Naval Research initiated a research effort in Non-Gaussian Signal Processing under the Selected Research Opportunities Program.

The program, sponsored by the Mathematical Sciences Division of ONR, brought together many of the leading researchers in statistics and signal processing. In addition to a consistent three year funding profile, a number of workshops and program reviews were held (Austin, Texas, October 1981; Annapolis, Maryland, May 1982; Kingston, Rhode Island, November 1982; Mt. Hood, Oregon, June 1983; and Princeton, New Jersey, March 1984) which brought together program investigators and interested Navy personnel from various facilities to exchange views and share the latest results of their research. This kind of basic research activity, motivated by real Navy problems in the non-Gaussian area, has not been previously undertaken in this country in such a focused manner. This book represents the culmination of the Non-Gaussian Program by reviewing the progress made in the program as well as presenting new results, not published elsewhere.

The book consists of fifteen papers, divided into three sections: Modeling and Characterization; Filtering, Estimation and Regression; Detection and Signal Extraction.

The first section starts with three papers which study and characterize a variety of ocean acoustic noise processes. The paper by Brockett, Hinich and Wilson looks at Gaussianity from a time series point of view and computes the bispectrum. Interestingly, time series which pass univariate tests of normality are shown, by means of bispectrum computation, to be non-Gaussian and non-linear time series. Powell and Wilson utilize the Middleton Class A model to characterize the noise from biological sources and acoustic energy generated by seismic exploration. They also investigate a number of techniques to estimate the parameters in the Class A canonical representation. In the study by Machell, Penrod and Ellis, kernel density estimates are obtained for the instantaneous amplitude fluctuations using five ocean ambient noise environments. Tests of homogeneity, randomness and normality are used to test the validity of the stationary Gaussian assumption. The fourth paper, by Mohler and Kolodziej, develops a model for a class of non-linear, non-Gaussian processes - the conditional linear and bilinear processes. They go on to identify distributions in this class which arise in underwater acoustical signal processing problems such as optimal state estimation. The final paper in this section by Wegman and Shull discusses a data analytic tool for displaying graphically the structure of multivariate statistical data. The procedure is called the parallel coordinate representation. Illustrative applications are given to multichannel time series, correlations and clustering, and beamforming for short segments of ocean acoustical data.

The second section of papers deals with filtering and estimation problems in a non-Gaussian environment. In the first study, Rosenblatt explores the structure of linear, non-Gaussian processes. It is shown that one can learn more about the linear structure from the observations than in the Gaussian case, a fact which should be helpful in modeling non-Gaussian autoregressive and moving average schemes. Rao investigates a class of nonstationary processes, the class which is subject to a generalized Fourier analysis. For this harmonizable class, one can use generalized spectral methods to do sampling, linear filtering, and signal extraction. The formulas for these operations, i.e., least squares estimation of harmonizable processes, have the same functional form as the results for stationary processes. Martin and Yohai study a robust estimate of ARMA model parameters called the AM-estimate. This estimate is based on a robust filter-cleaner which replaces outliers with interpolates based on previously cleaned data. Asymptotic properties are established and conditions for consistency are discussed. The paper by Brunk applies a Bayes least squares procedure and orthogonal transformation and expansion to the estimation of regression functions. Conditions for joint asymptotic normality of the transformed variables are established and an example is given which transforms a spectral density estimation problem into a regression problem.

The third section deals with the detection and extraction of signals. In the first paper, Bell considers the class of spherically exchangeable processes and treats five explicit signal detection models in this context. Both parametric and nonparametric techniques are utilized. The author points out that his procedures are generalizations of tests commonly used, that these tests are reasonable, but that little precise information is available on their performance. Next, Dwyer treats in considerable detail the development of processing techniques for the non-Gaussian noise encountered in the Arctic ocean and from helicopters. Both noise fields show significant narrow-band components. He proposes a method for the effective removal of these components to enhance signal detectability. In the third paper of this section, Tufts, Kirsteins, et al consider signal detection in a noise environment assumed to contain signal-like interference and impulse noise as well as a weak stationary Gaussian component. They develop iterative processing techniques to identify, categorize, model, and remove the non-Gaussian noise components, leaving a weak stationary Gaussian noise in which the weak signal, if present, may be detected. In another publication [Reference 6], they apply their methodology to Arctic undersea acoustic data. In the fourth paper, Baker presents a number of results in signal detection and communications. These include the detection of non-Gaussian signals in Gaussian noise and signal detection in non-Gaussian noise modeled as a spherically-invariant process. Both of these detection problems are immediately applicable to sonar. Continuing with the assumption of spherically-invariant processes, he considers the problem of mismatched channels and the capacity associated with spherically-invariant channels. The fifth paper, by Schwartz and Thomas, treats detection in a non-Gaussian environment where narrowband non-Gaussian components may be present and/or where the noise field may be nonstationary. In particular, they consider a switched detector based partly on the Middleton Class A noise model. They also consider the detection of fading narrowband signals in non-Gaussian noise using a robust signal estimator in conjunction with a robust detector. The last paper, by Machell and Penrose, considers energy detection in the context of ocean acoustic noise fields. Real noise data from this environment is used to evaluate the performance of energy detectors and it is found that their effectiveness depends heavily on the tail weights of the underlying noise distributions.

The editors would like to acknowledge the support and strong commitment to research of the Mathematical Sciences Division of the Office of Naval Research. We thank the contributors to this volume for the care in preparing their manuscripts and Karen Williams for her help in typing some of them. Finally, but most importantly, we are indebted to Diane Griffiths for the excellent job in preparing the manuscripts, often editing them so as to achieve uniformity in presentation.

CONTENTS

Preface v

Contributors xi

PART I: MODELING AND CHARACTERIZATION 1

1. Bispectral Characterization of Ocean Acoustic Time Series: Nonlinearity and Non-Gaussianity
P.L. Brockett, M. Hinich, and G.R. Wilson 2

2. Class A Modeling of Ocean Acoustic Noise Processes
D.R. Powell, and G.R. Wilson 17

3. Statistical Characteristics of Ocean Acoustic Noise Processes
F.W. Machell, C.S. Penrod, and G.E. Ellis 29

4. Conditionally Linear and Non-Gaussian Processes
R.R. Mohler and W.J. Kolodziej 58

5. A Graphical Tool for Distribution and Correlation Analysis of Multiple Time Series
E.J. Wegman and C. Shull 73

PART II: FILTERING, ESTIMATION AND REGRESSION 87

6. Comments on Structure and Estimation for NonGaussian Linear Processes
M. Rosenblatt 88

7. Harmonizable Signal Extraction, Filtering and Sampling
M.M. Rao 98

8. Fisher Consistency of AM-Estimates of the Autoregression Parameter Using Hard Rejection Filter Cleaners
R.D. Martin and V.J. Yohai 118

9. Bayes Least Squares Linear Regression is Asymptotically Full Bayes: Estimation of Spectral Densities
H.D. Brunk 128

PART III: DETECTION AND SIGNAL EXTRACTION 148

10. Signal Detection for Spherically Exchangeable (SE) Stochastic Processes
C.B. Bell 149

11. Contributions to Non-Gaussian Signal Processing
R.F. Dwyer 168

12. Detection of Signals in the Presence of Strong, Signal-Like Interference and Impulse Noise
D.W. Tufts, I.P. Kirsteins, P.F. Swaszek, A.J. Efron, and C.D. Melissinos 184

13. On NonGaussian Signal Detection and Channel Capacity
C. Baker 197

14. Detection in a Non-Gaussian Environment: Weak and Fading Narrowband Signals
S.C. Schwartz and J.B. Thomas 209

15. Energy Detection in the Ocean Acoustic Environment
F.W. Machell and C.S. Penrod 228

CONTRIBUTORS

C.R. BAKER, Department of Statistics, University of North Carolina, Chapel Hill, North Carolina

C.B. BELL, San Diego State University, La Jolla, California

P.L. BROCKETT, Department of Finance and Applied Research Laboratories, The University of Texas at Austin, Austin, Texas

H.D. BRUNK, Oregon State University, Corvallis, Oregon

R.F. DWYER, Naval Underwater Systems Center, New London, Connecticut

A.J. EFRON, Department of Electrical Engineering, University of Rhode Island, Kingston, Rhode Island

G.E. ELLIS, Applied Research Laboratories, The University of Texas at Austin, Austin, Texas

M. HINICH, Department of Government, The University of Texas at Austin, Austin, Texas

I.P. KIRSTEINS, Naval Underwater Systems Center, New London, Connecticut

W.J. KOLODZIEJ, Department of Electrical and Computer Engineering, Oregon State University, Corvallis, Oregon

F.W. MACHELL, Applied Research Laboratories, The University of Texas at Austin, Austin, Texas

R.D. MARTIN, The University of Washington, Seattle, Washington

C.D. MELISSINOS, Department of Electrical Engineering, University of Rhode Island, Kingston, Rhode Island

R.R. MOHLER, Department of Electrical and Computer Engineering, Oregon State University, Corvallis, Oregon

C.S. PENROD, The Applied Research Laboratories, The University of Texas at Austin, Austin, Texas

D.R. POWELL, The Applied Research Laboratories, The University of Texas at Austin, Austin, Texas

M.M. RAO, The University of California, Riverside, California

M. ROSENBLATT, University of California, San Diego, California

S.C. SCHWARTZ, Department of Electrical Engineering, Princeton University, Princeton, New Jersey

C. SCHULL, Department of Decision Sciences, University of Pennsylvania, Philadelphia, Pennsylvania

P.F. SWASZEK, Department of Electrical Engineering, University of Rhode Island, Kingston, Rhode Island

J.B. THOMAS, Department of Electrical Engineering, Princeton University, Princeton, New Jersey

D. TUFTS, University of Rhode Island, Kingston, Rhode Island

E.J. WEGMAN, Center for Computational Statistics and Probability, George Mason University, Fairfax, Virginia

G.R. WILSON, The Applied Research Laboratories, The University of Texas at Austin, Austin, Texas

V.J. YOHAI, Departmento de Matematica, Ciudad Universitaria, Buenos Aires, Argentina

PART I:

MODELING AND CHARACTERIZATION

BISPECTRAL CHARACTERIZATION OF OCEAN ACOUSTIC TIME SERIES: NONLINEARITY AND NON-GAUSSIANITY

Patrick L. Brockett*, Melvin Hinich**, Gary R.Wilson***
The University of Texas at Austin
Austin, Texas 78713-8029

ABSTRACT

Previous research into the Gaussianity of ocean acoustical time series has examined univariate marginal densities. In this paper we present research which examines this issue from a time series point of view. Even series which previously passed *univariate* tests for normality are shown to be non-Gaussian time series. Additionally, these time series are shown to be nonlinear time series, so that such acoustical series must be modeled in a nonlinear fashion.

I. INTRODUCTION

Optimal detection of signals in stochastic noise has been developed in only a relatively few cases. Most investigators assume that either the signal or the noise or both signal and noise are Gaussian processes. Even when non-Gaussianity is allowed, the process is often assumed to be a linear process, or a simple martingale derived from a Gaussian process (e.g., via simple stochastic differential equations). Examples of this include the results presented in Brockett (1984a,b) and Baker and Gaultierotti (1984). In many situations such as sonar, radar, or satellite transmission the Gaussianity assumption may not hold (c.f. Girodan and Haber (1972), Kennedy (1969), Trunk and George (1970), Van-Trees (1971), Wilson and Powell (1984), Machell and Penrod (1984), Milne and Ganton (1964), and Dwyer (1981)). It is thus of some interest to determine if signals and/or noise are Gaussian processes in the ocean environment. If they are non-Gaussian, we wish to determine if they can be modeled as non-Gaussian linear processes (e.g., autoregressive, autoregressive moving average, or simple stochastic integrals). This is useful for determining which type of signal detector to implement for best performance. If the series are both nonlinear and non-Gaussian, then new models must be developed.

Of course, in a time series setting one cannot detect nonlinearity or non-Gaussianity of a series by examination of the spectrum alone; hence, in this paper we shall examine the bispectrum of the acoustic time series as well. The technique we shall use to investigate the time series properties of the acoustical data is a statistical test for linearity and Gaussianity recently published by Hinich (1982) and based upon the bispectrum.

A summary of the paper is as follows: Section II presents necessary time series definitions and results; Section III presents a brief sketch of the Hinich statistical tests;

*Richard Seaver Centennial Fellow, IC^2 Institute, Department of Finance and Applied Research Laboratories, The University of Texas at Austin.

**Department of Government and Applied Research Laboratories, The University of Texas at Austin.

***Applied Research Laboratories, The University of Texas at Austin.

Section IV gives the results of applying these tests to acoustical time series; and Section V gives some directions for further research.

II. BACKGROUND

We shall first give some pertinent definitions and notation useful in the sequel for explaining our results.

A stationary time series $\{X(1), X(2),...\}$ is a sequence of random variables such that the joint distribution of $X(k_1),...,X(k_n)$ depends only on the difference between the k_i's and not on their precise values. Assuming the existence of moments, stationarity implies the mean $\mu = E[X(n)]$, the covariance $C_{xx}(m) = E[X(n+m)X(n)] - \mu^2$, and third moments $E[X(n+r)X(n+s)X(n)]$ are independent of n. All joint moments will be stationary as well. To simplify notation from here on we shall center our series and assume μ=0.

If $\{X(1), X(2),...\}$ are mutually independent, then the time series is called *purely random*. If $C_{xx}(m) = 0$ for all $m \neq 0$, then the series is called *white noise*. Purely random series are white, but not necessarily conversely. If the joint distribution of $\{X(k_1),...,X(k_n)\}$ is multivariate normal, then the time series is called a *Gaussian process*. White Gaussian processes are purely random, but in general whiteness of a series does not imply the series is purely random. This is an important distinction. Often researchers will make the assumption, for the sake of convenience, that a white residual series is Gaussian and will accept this normality as a fact when constructing detectors or doing hypothesis tests. If the series is non-Gaussian, this assumption may lead to erroneous inferences and suboptimal detectors. We should also note that the univariate marginal distribution of the series $X(t)$ may pass statistical tests for normality without the series being a Gaussian series. All joint distributions must also be normal before the Gaussianity of the series is assured.

A *linear process* is a time series which can be expressed in the form

$$X(n) = \sum_{m=-\infty}^{\infty} a(m)\, \epsilon(n-m) \tag{2.1}$$

where $\{\epsilon(n)\}$ is a purely random series. This model includes all the stable autoregressive, autoregressive moving average models, and certain martingale models which are often used. If the series $\{\epsilon(n)\}$ is Gaussian, then the original process $\{X(n)\}$ is also Gaussian. The converse is also true; $X(n)$ Gaussian implies $\epsilon(n)$ Gaussian, and all Gaussian processes are linear processes as well (i.e., (2.1) holds).

Standard tests for whiteness of a time series use the sample autocovariance function (or the power spectrum) of the series. If the series is, in fact, nonlinear, then the nonlinearities present when fitting a linear series model to the data will end up in the residual error term and will be undetected by standard tests based solely on the mean and covariance function. If the series is, in fact, nonlinear but the scientist uses only linear models for developing detection algorithms, then the performance of the detection may be much worse than is theoretically possible using the appropriate nonlinear model (even if the linear model is the "best fitting linear model").

In practice there are several ways in which a stationary non-Gaussian time series might arise. One method is that the series result from a nonlinear filtering of a Gaussian input process, i.e., a nonlinear filtering operation satisfying a Volterra functional expansion (c.f. Brillinger (1975) par. 2.10). Another method is that the process could have arisen as a linear filtering (e.g., ARIMA) model with a non-Gaussian input process. In this latter case the process is non-Gaussian but linear, while in the former case the process is both nonlinear and non-Gaussian. The tests presented subsequently allow us to differentiate between these two types of non-Gaussian processes.

III. STATISTICAL TESTS FOR LINEARITY AND GAUSSIANITY OF TIME SERIES

Subba Rao and Gabr (1980) and Hinich (1982) present statistical tests for determining whether a given stationary time series $\{X(n)\}$ is linear (i.e., has the form (2.1)) and Gaussian. It is possible that $\{X(n)\}$ is linear without being Gaussian but all the stationary Gaussian time series are linear.

Both the Subba Rao and Gabr and the Hinich tests are based upon the sample bispectrum of the time series. The Hinich test is nonparametric and is robust. Additionally, the Hinich test is conservative in the presence of certain types of nonstationarity of the time series (a frequent occurrence), so, if we can reject linearity and/or Gaussianity using the Hinich test, the rejection would probably continue even if the series was nonstationary. Accordingly, the tests presented in this paper use the Hinich test.

Let $\{X(n)\}$ be a mean zero stationary time series. The spectrum of $\{X(n)\}$ is the Fourier transform of the autocovariance function $C_{xx}(n) = E[X(t+n)\,X(t)]$.

The bispectrum of $\{X(n)\}$ is defined to be the (two dimensional) Fourier transform of the third moment function $C_{xxx}(n,m) = E[X(t+n)\,X(t+m)\,X(t)]$,

$$B(f_1,f_2) = \sum_m \sum_n C_{xxx}(n,m)\exp\{-\mathrm{i}f_1 n - \mathrm{i}f_2 m\}.$$

The principal domain of the bispectrum is the triangular set $\{0 \le f_1 \le \frac{1}{2}, f_2 \le f_1, 2f_1 + f_2 \le 1\}$ (Fig. 1). A rigorous introduction to the bispectra and their symmetries and properties can be found in Brillinger and Rosenblatt (1967). For our purposes in this paper the important thing about the bispectrum is that it allows a statistical test for linearity and Gaussianity of a time series.

By looking at the inverse transform we find that the third order cummulant C_{xxx} is the inverse transform of the bispectrum. In particular, at lags $m=n=0$,

$$E[X^3(t)] = C_{xxx}(0,0) = \int\int B(f_1,f_2)\,df_1\,df_2.$$

Thus the bispectrum can be interpreted as a frequency decomposition of the third order moment of the series, just as the spectrum can be interpreted as a frequency decomposition of the power σ_x^2 (second order cumulant) of the series. The bispectrum is a measure of frequency coherence of the series (c.f. Hasselman et al., 1963).

Suppose $X(n)$ is a linear time series, i.e., has the form (2.1). It can then be shown that the spectrum of $\{X(n)\}$ is of the form

$$S(f) = \sigma_\epsilon^2 \, |A(f)|^2 \tag{3.1}$$

and the bispectrum of $\{X(n)\}$ is of the form

$$B(f_1,f_2) = A(f_1)\, A(f_2)\, A^*(f_1 + f_2)\, \mu_3, \tag{3.2}$$

where $\mu_3 = E[\epsilon^3(t)]$, $A(f)$ is the transform of the coefficient series,

$$A(f) = \sum_{n=0}^{\infty} a(n)\, exp(-ifn)$$

and A^* is the complex conjugate of A.

From (3.1) and (3.2) it follows that

$$\frac{|B(f_1,f_2)|^2}{S(f_1)S(f_2)S(f_1+f_2)} = \frac{\mu_3^2}{\sigma_\epsilon^6} \tag{3.3}$$

is constant over all frequency pairs (f_1,f_2) if $\{X(n)\}$ is linear. Moreover, since $\mu_3 = 0$ for the normal distribution, the above constant is zero if the time series is Gaussian. (Of course, $\mu_3 = 0$ is possible for other non-Gaussian processes as well, but the point here is that $\mu_3 \neq 0$ is incompatible with Gaussianity.)

The relationship (3.3) is the basis of the Hinich tests. Constructing an estimate of the bispectrum $\hat{B}(f_1,f_2)$ and of the spectrum $\hat{S}(f)$, Hinich estimates the ratio in (3.3) at different frequency pairs (f_1,f_2) by $|\hat{B}(f_1,f_2)|^2/\hat{S}(f_1)\hat{S}(f_2)\hat{S}(f_1+f_2)$. If these ratios differ too greatly over different frequency pairs, he rejects the constancy of the ratio and hence linearity of the time series $\{X(n)\}$. If the estimates differ too greatly from zero, he rejects the Gaussianity time series model. The constant $\mu_3^2/\sigma_\epsilon^6$ is the square of Fisher's skewness measure for the ϵ series.

The test statistic Hinich derives for testing linearity is based upon the inner quartile range of the estimated ratio over the set of pertinent frequency pairs. If the ratio in (3.3) is constant, then the inner quartile range is small. If it is not constant, then the inner quartile range is larger. This test is robust and fairly powerful at sample sizes as low as 256 (c.f. Ashley, Hinich and Patterson (1986)). It is also conservative with respect to most types of nonstationarity since nonstationarity would tend to smear the peaks in the estimated bispectrum and hence reduce the dispersion of the estimated ratio. The test statistic for linearity is asymptotically normal, so significance is readily determined from standard normal tables. See Hinich (1982) for the precise formulae and proofs concerning this test for linearity.

The test for Gaussianity of the time series involves testing for the ratio (3.3) being zero. Hinich (1982) derives an asymptotically normal test statistic based upon the estimated ratio (3.3) in this situation as well.

IV. EMPIRICAL RESULTS FOR OCEAN ACOUSTIC DATA

The Hinich tests previously discussed were applied to several different data types: ambient noise (distant shipping and wind generated), individual ships, and biological noise (snapping shrimp). We wished to determine if the ambient noise (which passed certain univariate marginal tests for normality) was truly a linear Gaussian process, and also to characterize the time series behavior of certain potential signal-plus-noise processes. For detector construction this information and characterization is very useful.

For comparison purposes, the first series examined was a sequence of values generated by a random number generator that simulates Gaussian white noise. The series was divided into 100 records, each containing 1024 samples. Each record was tested for linearity and Gaussianity. The standardized, asymptotically normal test statistic values for the first 50 records are shown in Fig. 2 and are typical of all 100 records. The standardized sum of the linearity and Gaussianity test statistics are also shown in Fig. 2. Since the test statistics are one-sided, only large positive values are significant. This series can easily be accepted as linear and Gaussian.

The normalized bispectrum (3.3) for each record was averaged over all records. This average bispectrum is shown in Fig. 3. The bandwidth was arbitrarily set to 512 Hz. As is apparent, except for small random perturbations, the graph appears quite flat (as it should).

The next series examined was ocean ambient noise recorded in the northeast Pacific. The wind speed was 5-7 m/sec and the nearest known ship was over 500 km away. The power spectrum is shown in Fig. 4. No local shipping noise is evident. The spectrum below about 100 Hz is dominated by distant shipping, while the spectrum above about 100 Hz is primarily due to wind generated noise. The sharp falloff after about 500 Hz is due to low pass recording filters. The sampling rate for these data was 1500 Hz.

This ambient noise time series was divided into 200 records, each containing 1024 samples, and the Hinich tests for linearity and Gaussianity were applied to each record. The standardized test statistic values for the first 50 records are shown in Fig. 5 and are typical of all 200 records.

The sum statistic should again be compared with a standard normal variate for statistical inference. Using the inequality

$$\frac{1}{\sqrt{2\pi}}\{\frac{1}{x}-\frac{1}{x^3}\}\exp\{-\frac{x^2}{2}\} < 1-\Phi(x) < \frac{1}{x\sqrt{2\pi}}\exp\{\frac{-x^2}{2}\},$$

where Φ is the standard normal distribution function, we may easily assess the probability that the time series is truly Gaussian (p-value of 4.5×10^{-20}) or truly linear (p-value of 3×10^{-6}). Thus we observe that this series is virtually incompatible with either a linear process model or a Gaussian process model. Of particular note is that this series easily passed tests for univariate normality.

Figure 6a shows a plot of the normalized bispectrum for this series averaged over the first 100 records. Figure 6b shows the average of the next 100 records. The spectrum was computed for the first 60 records and averaged to give the spectrum $S(f)$ used to compute the normalized bispectrum for each of the 200 records. As is apparent, the normalized bispectrum for this real acoustical data is quite consistent over time, and quite visually distinct from the flat surface which would be obtained from a linear time series. Moreover, there are distinct peaks observable at the lower frequency pairs which account in part for this nonlinearity.

The third ocean acoustic data set considered was dominated by a single merchant ship crossing a sensor. Previous studies had shown that the marginal densities of this series were significantly non-Gaussian (Machell and Penrod (1984)); however, the linearity (e.g., perhaps autoregressive or autoregressive moving average) had not been previously examined. The sampling rate for this series was 1250 samples per second. Two hundred records of length 1024 were taken, and the spectrum was estimated by averaging the first 60 records. This estimated spectrum was then used to normalize the bispectral estimate in each record. This power spectrum is shown in Fig. 7. The periodic component of the power spectrum is indicative of the acoustical energy from the merchant ship arriving over more than one path. In this data set, even the high frequency energy is dominated by the merchant.

Figure 8 shows the individual test statistic values for each of the first 50 records. Additionally, the sum statistic obtained by averaging all 50 records is presented. All values here should again be compared with the standard normal distribution for significance testing. We find the series to be significantly non-Gaussian (p-value $< 3.5 \times 10^{-113}$) and also nonlinear (p-value $< 1.7 \times 10^{-18}$).

Figures 9a and 9b show the plotted values of the normalized bispectrum for this data averaged over two consecutive sets of 100 records each. Again, it is apparent that the plotted curve is non-flat. The peaks appearing in the normalized bispectrum are consistent over time although their level is somewhat lower in the second set of records. An interesting observation is that the plotted bispectrum, while consistent for this merchant ship, is quite visually distinct from the ambient noise plot (Fig. 6).

Some of the peaks in this bispectrum estimate may be an anomaly of the estimation procedure that is particularly significant with this data set. Figure 10 shows the same bispectrum as shown in Fig. 9a, but in a contour format. The frequencies at which the peaks occur are easier to observe in this format. As can be seen, most of the peaks at the higher frequencies occur at pairs of frequencies corresponding to local minima in the power spectrum (Fig. 7). One possible explanation may be as follows. To achieve consistency, the bispectrum estimate is averaged over a square of frequency components about a pair of frequencies, where the averaging length is greater than the square root of the number of samples in the Fourier transform of the time series. However the power spectrum was not averaged over the range of frequency components comprising each square. If the bispectrum and power spectrum are slowly varying over this range of frequencies, then the normalized bispectrum has little bias for the transform lengths used here. In this particular case the averaging length was about 40 Hz. It can be seen from Fig. 7 that the power spectrum can change significantly over a 40 Hz range of frequencies. Thus this estimation procedure may not accurately represent the bispectrum at the higher frequencies. In particular, at the frequencies corresponding to the minima of the power spectrum, the normalized bispectrum may be overestimated.

The final acoustic data set to be considered is biologically generated noise (snapping shrimp). Previously this was also determined to have non-Gaussian marginals (Wilson and Powell (1984)), and this fact is collaborated by the Hinich test statistic values calculated in Fig. 11. Again, one rejects both Gaussianity and linearity of the observed time series data.

V. CONCLUSIONS AND EXTENSIONS

We have shown that with new time series tests we can determine that some series which were previously believed to be Gaussian (such as ambient noise) are in fact non-linear and non-Gaussian. Further, by examining $\hat{B}(f_1,f_2)/\hat{S}(f_1)\hat{S}(f_1)\hat{S}(f_2)\hat{S}(f_1+f_2)$ we can determine precisely which frequency pairs have significant coupling. The nonlinearity of such series was also shown to be a rather consistent result in ocean acoustic data.

Several important extensions and applications of these results are being considered. One application of these techniques is to test the Middleton reverberation (linear) model (Middleton (1967a,b, 1972a,b)) using real data. This has not been examined previously in this context. These techniques can also be applied to characterizations of the ambient noise series in various different ocean environments. This characterization could be useful for detector development adapted to specific geographical regions.

The striking dissimilarities between the normalized bispectrum of different noise sources (Figs. 3, 6, and 9) and the similarities apparent between the normalized bispectrum of a single source over time motivates the subsequent examination of whether bispectral analysis can be used as a source identification procedure. This is currently under study.

Theoretical extensions are also under consideration. One such extension is the use of kernel estimation procedures for obtaining improved bispectral density estimators. The multivariate structure of the resulting bispectrum will be developed, and the trispectrum and its properties will also be developed. The difficulties associated with the bispectrum estimate of the merchant demonstrate the need to develop averaging procedures that do not require that the bispectrum and power spectrum be slowly varying. The convergence of the bispectrum estimate, as it is averaged over several records, must also be examined.

REFERENCES

1. Ashley, R.A., M.J. Hinich and D. Patterson, "A Diagnostic Test for Non-Linear Serial Dependence in Time Series Fitting Errors," *J. of Time Series Analysis,* Vol. 7, No. 3, pp. 165-178, 1986.

2. Baker, C.R., and A.F. Gualtierotti, "Detection of Non-Gaussian Signals in Gaussian Noise," Statistical Signal Processing, edited by E.J. Wegman and J.G. Smith, Marcel Dekker, Inc., New York, pp.107-116, 1984.

3. Brillinger, D., Time Series, Data Analysis and Theory, Holt, Rinehart, and Winston, New York, 1975.

4. Brillinger, D. and M. Rosenblatt, "Asymptotic Theory of kth Order Spectra," Spectral Analysis of Time Series, pp. 153-158.

5. Brockett, P.L., "Optimal Detection in Linear Reverberation Noise," Statistical Signal Processing, edited by E.J. Wegman and J.G. Smith, Marcel Dekker, Inc., New York, pp. 133-140, 1984a.

6. Brockett, P.L., "The Likelihood Ratio Detector for Non-Gaussian Infinitely Divisible and Linear Stochastic Processes," *Annals of Statistics,* Vol. 12, No. 2, pp. 737-74, 1984b.

7. Dwyer, R.F., "FRAM II Single Channel Ambient Noise Statistics," NUSC Rpt. No. TD 6583, Naval Underwater Systems Center, New London, Connecticut, 1981.

8. Girodan, A.A., and F. Haber, "Modeling of Atmospheric Noise," *Radio Science,* Vol. 7, pp. 1011-1023, 1972.

9. Hasselman, K.W. Munk, and G. MacDonald, "Bispectra of Ocean Waves," *Proc. Symp. Time Series Analysis,* edited by M. Rosenblatt, Wiley, New York, pp. 135-13, 1963.

10. Hinich, M., "Testing for Gaussianity and Linearity of a Stationary Time Series," *Journal Time Series Analysis,* Vol. 3, No. 3, pp. 169-176, 1982.

11. Kennedy, R.S., Fading Dispersive Communication Channels, Wiley, New York, 1969.

12. Machell, F.W., and C.S. Penrod, "Probability Density Functions of Ocean Acoustic Noise Processes," Statistical Signal Processing, edited by E.J. Wegman and J.G. Smith, Marcel Dekker, Inc., New York and Basel, pp. 211-221, 1984.

13. Middleton, D., "A Statistical Theory of Reverberation and Similar First Order Scattered Fields. Part I: Waveform and General Processes," *IEEE Trans. Inf. Theory,* Vol. 13, pp. 372-392, 1967a.

14. Middleton, D., "A Statistical Theory of Reverberation and Similar First Order Scattered Fields, Part II: Moments, Spectra, and Spatial Distributions," *IEEE Trans. Inf. Theory,* Vol. 13, pp. 393-414, 1967b.

15. Middleton, D., "A Statistical Theory of Reverberation and Similar First Order Scattered Fields. Part IV: Statistical Model," *IEEE Trans. Inf. Theory,* Vol. 18, pp. 35-67, 1972a.

16. Middleton, D., "A Statistical Theory of Reverberation and Similar First Order Scattered Fields. Part IV: Statistical Models," *IEEE Trans. Inf. Theory,* Vol. 18, pp. 68-90, 1972b.

17. Milne, A.R., and J.H. Ganton, "Ambient Noise under Arctic Sea Ice," *J. Acoust. Soc. Am.,* Vol. 36, pp. 855-863, 1964.

18. Suba Rao, T., and M.M. Gabr, "A Test for Linearity of Stationary Time Series," *Journal Time Series Analysis,* Vol. 2, pp. 145-158.

19. Trunk, G.V., and S.F. George, "Detection of Targets in Non-Gaussian Sea Clutter," *IEEE Trans. Aerosp. Electron. Syst.,* Vol. AES-6, pp. 620-628, 1970.

20. VanTrees, H.L., Detection, Estimation, and Modulation Theory, Part III: Radar–Sonar Signal Processing and Gaussian Signals in Noise, Wiley, New York, 1971.

21. Wilson, G.R., and D.R. Powell, "Experimental and Modeled Density Estimates of Underwater Acoustic Returns," in Statistical Signal Processing, edited by E.J. Wegman and J.G. Smith, Marcel Dekker, Inc., New York and Basel, pp. 223-239, 1984.

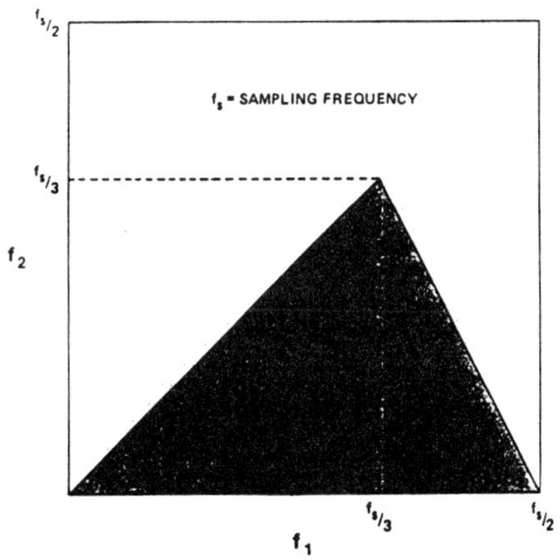

Figure 1 Principal domain of the bispectrum.

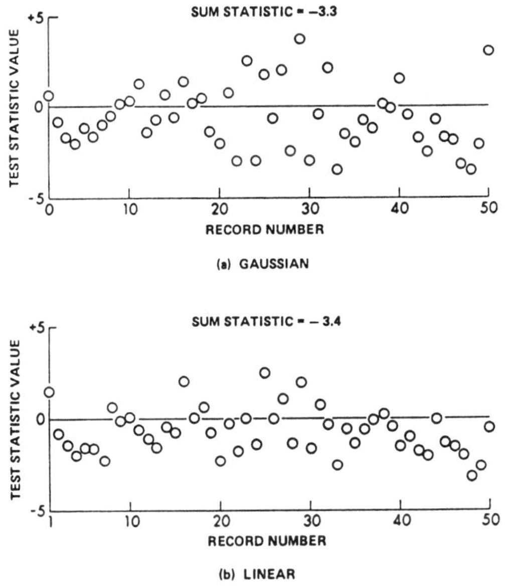

Figure 2 Linear and Gaussian test statistic values - Gaussian random numbers.

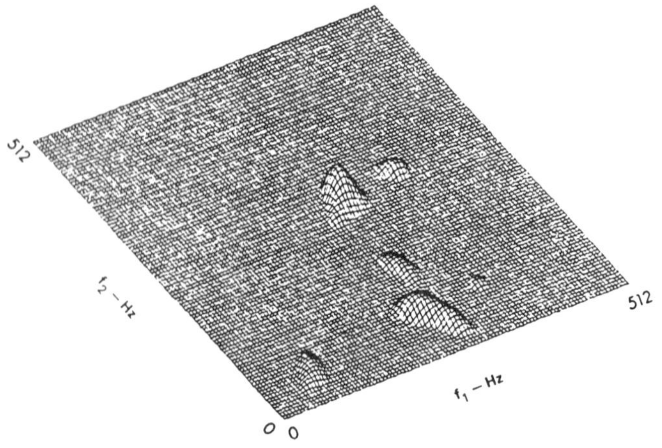

Figure 3 Average normalized bispectrum - Gaussian random numbers.

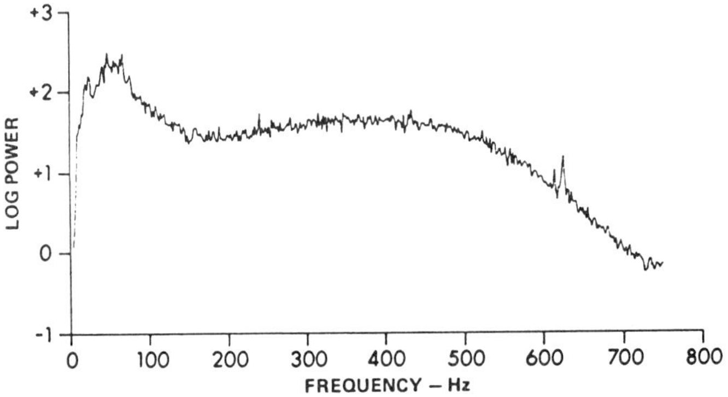

Figure 4 Power spectrum - deep ocean ambient noise.

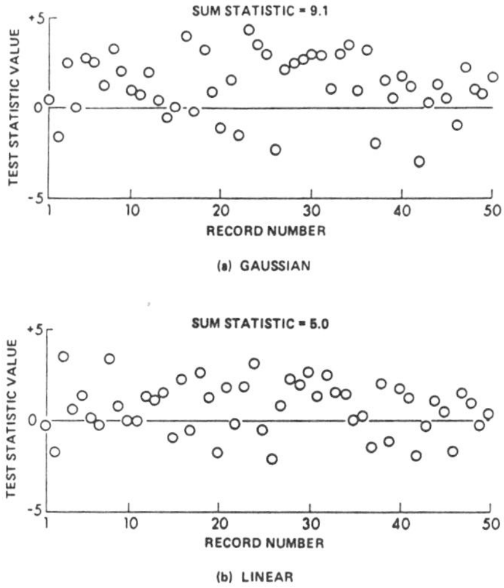

Figure 5 Linear and Gaussian test statistic values - deep ocean ambient noise.

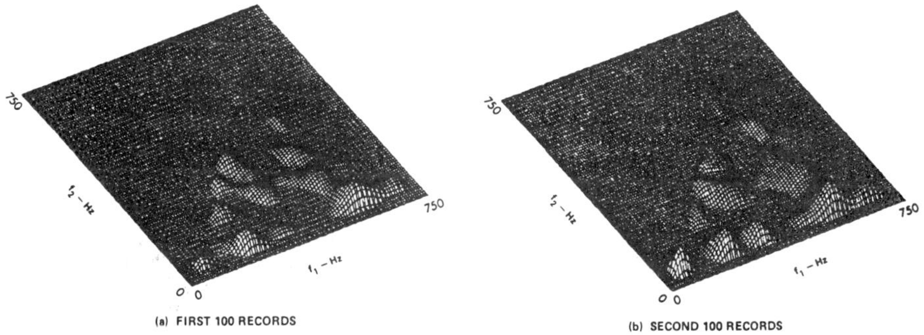

Figure 6 Average normalized bispectrum - deep ocean ambient noise
(a) First 100 records
(b) Second 100 records

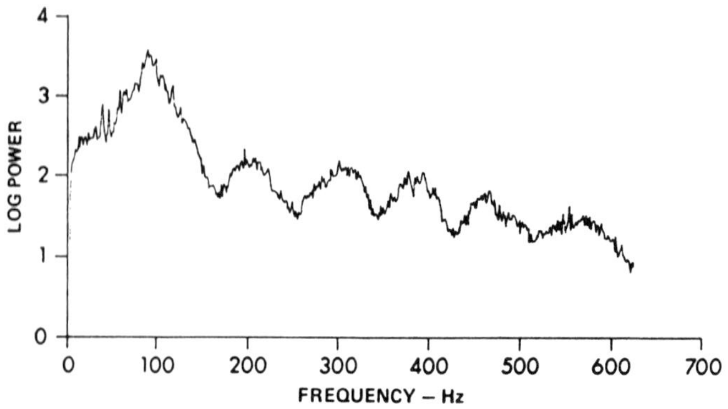

Figure 7 Power spectrum - merchant radiated noise.

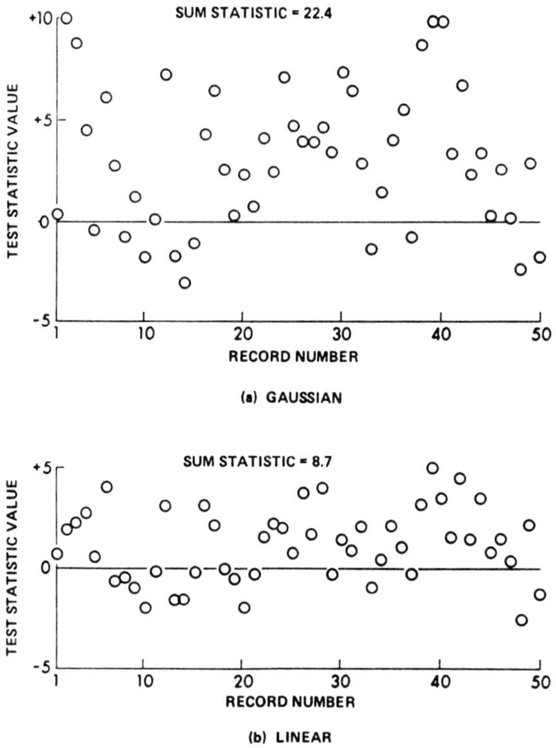

Figure 8 Linear and Gaussian test statistic values - merchant radiated noise.

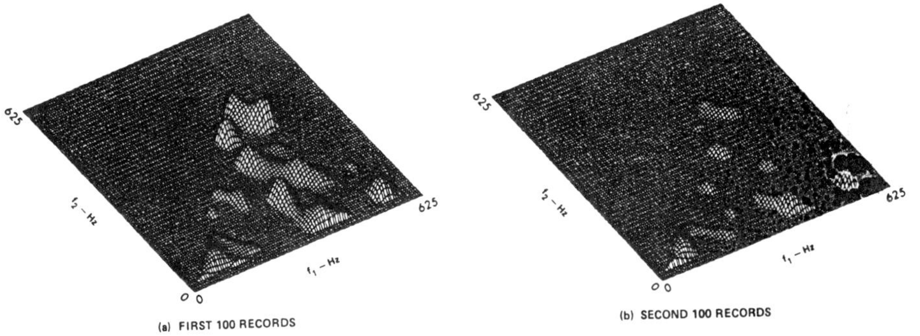

Figure 9 Average normalized bispectrum - merchant radiated noise
(a) First 100 records
(b) Second 100 records

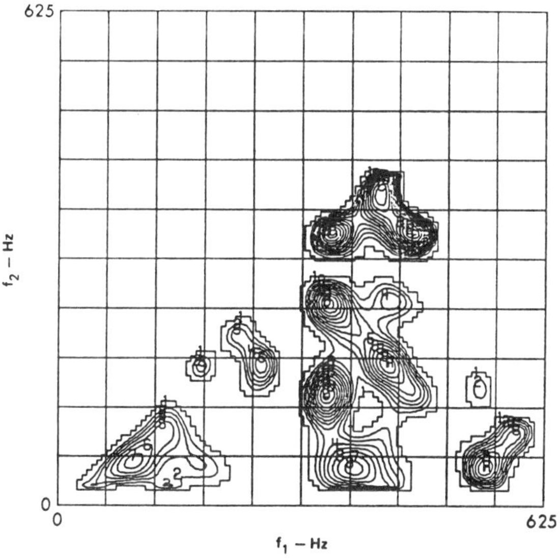

Figure 10 Average normalized bispectrum - merchant radiated noise

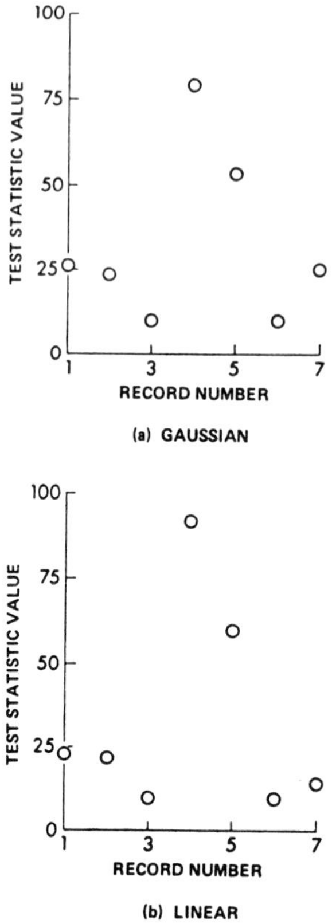

Figure 11 Linear and Gaussian test statistic values - biological noise

CLASS A MODELING OF OCEAN ACOUSTIC NOISE PROCESSES

Dennis R. Powell and Gary R. Wilson
Applied Research Laboratories
The University of Texas at Austin

ABSTRACT

Previous work has shown that some ocean acoustic noise processes can be represented as Class A noise. Likelihood ratio and threshold detectors have been developed to detect signals in the presence of Class A noise. The performance of these detectors is significantly affected by the accuracy with which the parameters of the Class A noise can be estimated. This paper presents two methods of estimating the Class A parameters, a minimum distance method and a maximum likelihood method. These methods are compared to a previously developed method using estimates of the moments of the noise process and are generally found to be superior estimators.

I. INTRODUCTION

Acoustic signal processing in an ocean environment is often performed under the assumption that the probability density functions of the signal and/or noise processes are from a certain class, usually Gaussian. This assumption leads to analytically tractable processing algorithms, and in many instances it is a reasonable assumption to make. However, it is known that some underwater acoustic noise processes are non-Gaussian.[1–5] Two different processing approaches can be taken in these non-Gaussian environments where the true governing probability laws are perhaps unknown. One approach is a robust, nonparametric approach, which does not require an exact knowledge of the density function of the process, but which requires perhaps less specific information about the process.[6–11] This approach may involve either adaptive or non-adaptive processors. The other approach is a parametric approach, which usually requires knowledge of the classes of density functions describing the signal and noise processes. A generalized likelihood ratio would be an example of the latter approach. In some cases a likelihood ratio can be implemented without a precise knowledge of the density function.[12–13]

One difficulty with the parametric approach is the problem of estimating the probability density function when it is not known a priori. Various nonparametric techniques have been developed to estimate the probability density function from observations of the random process.[14–16] These techniques assume little or no prior information about the process and provide only gross information to guide the identification of an appropriate parametric class of densities. Alternatively, physical models of non-Gaussian processes have been developed that allow specification of the density function of the process based on a few physically motivated parameters. The most general of these models are those developed by Middleton.[17] Likelihood ratio and threshold detectors have been developed for detecting signals in the presence of non-Gaussian noise which can be characterized by one of Middleton's noise models.[18–20] However, the performance of these detectors is critically dependent on the accuracy of the parameters that characterize the model. Various methods have been proposed for estimating these parameters for the Middleton Class A model,[21] and these methods have been applied to experimental underwater acoustic data.[1] The purpose of this paper is to present and compare some additional methods for estimating these parameters. The comparison will be performed using two different sets of underwater acoustic data. First the results of various statistical tests for randomness, homogeneity, and univariate normality of the data will be presented. Then the Middleton Class A noise model and the parameter estimation methods will be described. A comparison of the methods will be made and some conclusions will be drawn.

II. DATA VALIDATION AND CHARACTERIZATION

The two data sets that are used in this paper are generated by significantly different mechanisms. One data set, referred to as the seismic data, represents acoustic energy generated by impulsive sources used for seismic exploration. The other data set is referred to as the biological data and represents acoustic energy generated from a biological source, most likely snapping shrimp. Both data sets were collected in the Gulf of Mexico, although on different occasions. Each data set was digitized, and 1000 samples of the time series were extracted from each set for use in this analysis. The time series of each set is shown in Fig. 1.

Both data sets were tested for randomness and homogeneity. The runs up and down test was used to test for randomness, while the Kolmogorov-Smirnov two-sample test and the Wilcoxon rank sum test were used to test for homogeneity. Descriptions of these tests can be found elsewhere.[22–23] Table I gives the two-sided p-values for each data set and test. A p-value is the probability of exceeding the test statistic value that resulted from the test if the null hypothesis were true. A small p-value (less than 0.1, for example) indicates that the null hypothesis of the test can be rejected with some confidence. As can be seen from Table I, none of the p-values are small, indicating that both of the data sets can be accepted as random and homogeneous.

Univariate tests for normality were also applied to both data sets. The tests that were used were Pearson's tests of skewness and kurtosis, and D'Agostino's test.[22–23] The p-values for these tests are shown in Table II. The low p-values indicate that both data sets can be considered with a reasonable degree of confidence to be non-Gaussian. The skew and kurtosis for each data set are shown in Table III. They differ significantly from the Gaussian values of 0 and 3, respectively. The larger than normal kurtosis indicates a heavy-tailed distribution.

III. CLASS A MODEL

If the Gaussian hypothesis is not supported by statistical tests of normality, alternative probability models must be considered. Both parametric and nonparametric models are available. Nonparametric methods for density estimation utilize no distributional assumptions and estimate the density directly from the data. Typically, the form of the estimate is either a series of kernel functions[14–15] or a series of orthogonal functions that satisfy the conditions of a probability density function.[16] In the first method the density estimate is given by

$$\hat{f}_n(x) = \frac{1}{n} \sum_{i=1}^{n} \frac{1}{h(n)} K(\frac{x - X_i}{h(n)})$$

where $K(x)$ is a kernel function, $h(n)$ defines a sequence of kernel widths, and $X_1,\ldots,X_n$ are the observed data. Often $h(n)$ is a constant with its optimal value proportional to $n^{-0.2}$, with the constant of proportionality dependent upon the true underlying density. Examples of kernel density estimates are given in Figs. 2-5. Nonparametric methods can provide computationally fast univariate density estimates that indicate the shape of the underlying probability density function (pdf) and may give insight into the non-Gaussian nature of the data. On the other hand, signal detection theory is not well developed for density estimates which cannot be parameterized by a small number of variables.

A useful parametric model of non-Gaussian acoustic processes is Middleton's Class A model.[17,21] This model is based on Middleton's theory of impulsive interference for the special case where the interference has bandwidth less than the receiver. The density is given in standardized form as an infinite mixture of normals:

$$\varsigma(z;A,\Gamma) = e^{-A} \sum_{m=0}^{\infty} \frac{A^m}{m!} \phi(z,0,\sigma_m^2) \quad A > 0, \ \Gamma > 0$$

where

$$\sigma_m^2 = (\frac{m}{A} + \Gamma)/(1 + \Gamma)$$

and $\phi(z;\mu,\sigma^2)$ is the normal probability density with mean μ and variance σ^2.

The Middleton Class A density is completely characterized by the parameters A and Γ. The A parameter is a measure of the impulsiveness of the interference and may be interpreted as the product of the average number of impulses emitted per unit of time and the average duration of an impulse. Small values of A indicate a highly impulsive process while large values indicate a less impulsive, nearly Gaussian process. The Γ parameter is the ratio of the intensity of the independent Gaussian component of the process to the impulsive component.

Given an experimental acoustic data set $\{X_1,X_2, \ldots, X_n\}$, several methods exist to estimate the model parameters: method of moments (MOM), minimum distance (MD), and maximum likelihood (ML). The MOM estimates may be derived from the characteristic function of the Class A density,[21] yielding

$$\hat{A}_{MOM} = \frac{225(m_4 - 3)^3}{27(m_6 - 15m_4 + 30)^2};$$

$$\hat{\Gamma}_{MOM} = \frac{9(m_6 - 15m_4 + 30)}{15(m_4 - 3)^2},$$

where $m_j = \frac{1}{n} \sum_{i=1}^{n} Z_i^j$ and Z_i is X_i standardized to zero mean and unit variance.

An advantage the MOM has, other than computational efficiency, is consistency, i.e., the estimates converge in probability to the true parameter values. However, the MOM algorithm can readily yield negative values for A or Γ for sample sizes as large as several thousand. These invalid estimates (A and Γ must be greater than zero) occur frequently in our experience, 50-80% of the time. Thus, convergence of the MOM estimators for the Class A model must be considered to be "slow."

The MD method is a heuristic technique which matches the observed empirical distribution function (EDF) to the model density.[24] The EDF is defined as

$$\eta(x) = \frac{k}{n},$$

where

$$X_{(k)} < x \le X_{(k+1)}$$

and $X_{(k)}$ is the kth order statistic, i.e., $X_{(1)} \le X_{(2)} \le \cdots \le X_{(n)}$.

Given a probability density model ψ with parameters $\underset{\rightarrow}{\theta} = (\theta_1, \theta_2, \ldots, \theta_m)$, then

$$L_p(\eta, \underset{\rightarrow}{\theta}) = \int_{-\infty}^{\infty} \left| \int_{-\infty}^{x} \psi(t, \underset{\rightarrow}{\theta})dt - \eta(x) \right|^p dx, \quad p > 0.$$

If L is minimized over $\underset{\rightarrow}{\theta}$, then the value of $\hat{\theta}_{\rightarrow MD}$ which achieves the minimum is the MD estimator. Under mild regularity conditions for the underlying density, MD estimators are consistent.[25] Often L is analytically intractable and numerical techniques are utilized to derive estimates. Various distance measures have been proposed other than L_p norms. In this paper, a value of p=1 was used to generate the MD estimates. MD estimates provide robust performance even in cases where the hypothesized model is incorrect. Thus, the MD method is particularly appropriate when the true underlying density is unknown but a "reasonable" model is assumed.

The maximum likelihood method is well known. Given a density $\psi(x, \underset{\rightarrow}{\theta})$ and a random sample $\underset{\sim}{X} = \{X_1, \ldots, X_n\}$, the likelihood function is $\lambda = \prod_{i=1}^{n} \psi(X_i, \underset{\rightarrow}{\theta})$, and the log likelihood is

$$\Lambda = \ln\lambda = \sum_{i=1}^{n} \ln \psi(X_i, \underset{\rightarrow}{\theta}).$$

The parameters $\hat{\theta}_{\rightarrow ML}$ which maximize Λ are the maximum likelihood estimates. For many distributions the ML estimates are analytically derived by solving the homogeneous system of differential equations $\frac{\partial \Lambda}{\partial \theta_j} = 0$, $j = 1,2,\ldots,m$. However, an analytical solution has not been derived for the Middleton Class A density. Again, numerical techniques can be employed to approximate the ML estimates.

Previous analysis of non-Gaussian acoustic data led to the attempt to model the data with the Middleton Class A density.[1] Only the MOM estimate was utilized, providing initially unsuitable results. An acceptable parameter set was eventually derived but only after additional manipulation. The typical problem arising in the MOM estimates was the occurrence of invalid parameter estimates, i.e., $\hat{A}$ or $\hat{\Gamma}$ would have a value less than zero, violating the parameter restrictions of $A > 0, \Gamma > 0$ in the definition of the Class A density. Alternate estimators were sought to circumvent this problem.

The MD method was chosen as a candidate estimator for its robust behavior and its heuristic appeal. The ML was also chosen for its general acceptance as a desirable estimator. As applied to the Class A density, neither estimation technique admits an analytic solution (at least none discovered by the authors). Numerical methods were applied to approximate the MD and ML estimators. To assess the relative performance of the MOM, MD, and ML methods, data sets were randomly generated from a Class A distribution with parameters $A = 0.2, \Gamma = 1.0$ (kurtosis of 6.75). Thirty data sets were generated, ten each of size 50, 100, and 300. Estimates of bias, variance, and mean squared error (MSE) were computed for each parameter using the formulas:

$$b(\theta) = \frac{1}{K}\sum_{i=1}^{K}\hat{\theta}_i - \theta = \bar{\theta} - \theta$$

$$Var(\theta) = \frac{1}{K}\sum_{i=1}^{K}(\hat{\theta}_i - \bar{\theta})^2$$

$$MSE(\theta) = Var(\theta) + [b(\theta)]^2,$$

where $\hat{\theta}_i$ is the estimate of A or Γ for the ith data set and $K = 10$. The results are given in Table IV. These Monte Carlo results can provide only a rough indication of the performance of the estimation methods since the number of estimates for each sample size is small ($K = 10$). It can be concluded that both the MD and ML methods are superior to MOM, although sufficient variation exists in the MD and ML performance to preclude any conclusion that one is better than the other. More extensive simulation is required to confidently determine the superior method.

Of the three methods, the MD and ML methods are the more computationally intensive. For very large sample sizes (greater than several thousand) computational time may become prohibitive. On a CDC CYBER 171 computer the MD algorithm required 4500 times more CPU time than the MOM method, while ML required 17000 times more CPU time.

The MD and ML estimation procedures were applied to the snapping shrimp and seismic data described in the previous section. The Class A densities derived by these methods are compared to kernel density estimates in Figs. 2-5. Fixed width kernels were computed with $h = 0.3\hat{\sigma}$, where $\hat{\sigma}$ is the sample standard deviation. Figure 2 compares the Class A density with parameters estimated by MD to the kernel density for the shrimp data. The densities seem to be in good agreement, except at the cusp, with reasonable agreement in the tails. Figure 3 compares the kernel estimate with the ML Class A density derived from the shrimp data. Greater agreement is observed in the cusp and although the tail behavior is somewhat different, it remains reasonable based on a comparison with this kernel estimate. A kernel estimate of the snapping shrimp data has been previously made using a much larger sample size.[1] Based on that density estimate, which gives a better indication of the tail behavior, the Class A model derived from the MD estimator appears to better represent the tail behavior than the ML estimator. Figures 4 and 5 show the MD and ML Class A models, respectively, compared to the kernel estimate derived from the seismic data. Comments given for the shrimp densities also apply to the seismic data. The MD estimators yield a model that underestimates the cusp but otherwise provides reasonable agreement. The ML estimators tend to provide better agreement in the cusp and maintain reasonable estimates for tail behavior.

IV. CONCLUSIONS

The ML estimator appears to provide a better estimate of the density in the cusp than the MD estimator and provides acceptable estimates in the tails if the density is not too heavy-tailed. However, the MD estimator appears to better represent the tails of a heavy-tailed density such as that of the snapping shrimp. Additional estimates of the bias, variance, and MSE of the estimators for a variety of values for A and Γ will be necessary to confirm this speculation. It may be necessary in practice to have some estimate of the tail behavior before deciding which estimator to use.

In any case, both the ML and MD estimators appear to be reasonable candidates for estimating the Class A parameters and are generally superior to the moment estimator. These estimators can be used in conjunction with the likelihood ratio and threshold detectors already developed for Class A noise to provide an adaptive parametric technique for the detection of signals in non-Gaussian noise.

ACKNOWLEDGMENTS

This work was supported under the Probability and Statistics Program with the Office of Naval Research. Parts of this work were performed with student programming support under General Research Program, Support of High School Students and Supervisors, with the Office of Naval Research.

REFERENCES

1. G.R. Wilson and D.R. Powell, "Experimental and Modeled Density Estimates of Underwater Acoustic Returns," in Statistical Signal Processing, edited by E.J. Wegman and J.G. Smith, Marcel Dekker, Inc., New York and Basel, pp. 223-239, 1984.

2. F.W. Machell and C.S. Penrod, "Probability Density Functions of Ocean Acoustic Noise Processes," in Statistical Signal Processing, edited by E.J. Wegman and J.G. Smith, Marcel Dekker, Inc., New York and Basel, pp. 211-221, 1984.

3. G.R. Wilson and D.R. Powell, "Probability Density Estimates of Surface and Bottom Reverberation," *J. Acoust. Soc. Am.,* Vol. 73, No. 1, pp. 195-200, 1983.

4. A.R. Milne and J.H. Ganton, "Ambient Noise under Arctic Sea Ice," *J. Acoust. Soc. Am.,* Vol. 36, pp. 855-863, 1964.

5. R.F. Dwyer, "FRAM II Single Channel Ambient Noise Statistics," NUSC Rpt. No. TD 6583, Naval Underwater Systems Center, New London, Connecticut, 1981.

6. J.B. Thomas, "Nonparametric Detection," *Proc. IEEE,* Vol. 58, pp. 623-631, 1970.

7. R.D. Martin and S.C. Schwartz, "Robust Detection of a Known Signal in Nearly Gaussian Noise," *IEEE Trans. Inf. Theory,* Vol. IT-17, pp. 50-56, 1971.

8. J.H. Miller and J.B. Thomas, "Robust Detectors for Signals in Non-Gaussian Noise," *IEEE Trans. Commun.,* Vol. COM-25, No. 7, pp. 686-690, July 1977.

9. E.J. Modugno and J.B. Thomas, "The Detection of Signals in Impulsive Noise," Report No. 13, Information Sciences and Systems Laboratory, Princeton University, Princeton, New Jersey, June 1983.

10. S.V. Czarnecki and J.B. Thomas, "Nearly Optimal Detection of Signals in Non-Gaussian Noise," Report No. 14, Information Sciences and Systems Laboratory, Princeton University, Princeton, New Jersey, February 1984.

11. S.C. Schwartz and J.B. Thomas, "Detection in a Non-Gaussian Environment", Statistical Signal Processing, edited by E.J. Wegman and J.G. Smith, Marcel Dekker, Inc., New York and Basel, pp. 93-105, 1984.

12. C.R. Baker and A.F. Gualtierotti, "Detection of Non-Gaussian Signals in Gaussian Noise," Statistical Signal Processing, edited by E.J. Wegman and J.G. Smith, Marcel Dekker, Inc., New York and Basel, pp. 107-116, 1984.

13. P.L. Brockett, "Optimal Detection in Linear Reverberation Noise," Statistical Signal Processing, edited by E.J. Wegman and J.G. Smith, Marcel Dekker, Inc., New York and Basel, pp.133-140, 1984.

14. M. Rosenblatt, "Remarks on Some Nonparametric Estimates of a Density Function," *Ann. Math. Stat.*, Vol. 27, pp. 832-837, 1956.

15. E. Parzen, "On Estimation of a Probability Density Function and Mode," *Ann. Math. Stat.*, Vol. 33, pp. 1065-1076, 1962.

16. M.E. Tartar and R.A. Kronmal, "An Introduction to the Implementation and Theory of Nonparametric Density Estimation," *The American Statistician,* Vol. 30, pp. 205-212, 1976.

17. D. Middleton, "Statistical-Physical Models of Electromagnetic Interference," *IEEE Trans. Electromagn. Compat.*, Vol. EMC-19, pp. 106-127, 1977.

18. A.D. Spaulding and D. Middleton, "Optimum Reception in an Impulsive Interference Environment - Part I: Coherent Detection," *IEEE Trans. Commun.*, Vol. COM-25, pp. 910-923, 1977.

19. K.S. Vastola and S.C. Schwartz, "Suboptimal Threshold Detection in Narrowband Non-Gaussian Noise," presented at the *1983 IEEE International Conference on Communications,* Boston, Massachusetts, June 1983. To be published in the proceedings of the conference.

20. S.C. Schwartz and K.S. Vastola, "Detection of Stochastic Signals in Narrowband Non-Gaussian Noise," *Proceedings of the 22nd IEEE Conference on Decision and Control,* pp. 1106-1109, December 1983.

21. D. Middleton, "Procedures for Determining the Parameters of the First-Order Canonical Models of Class A and Class B Electromagnetic Interference," *IEEE Trans. Electromagn. Compat.* Vol. EMC-21, pp. 190-108, 1979.

22. G.R. Wilson, "Covariance Functions and Related Statistical Properties of Acoustic Backscattering from a Randomly Rough Air-Water Interface," Applied Research Laboratories Technical Report No. 81-23 (ARL-TR-81-23), Applied Research Laboratories, The University of Texas at Austin, 19 June 1981.

23. G.R. Wilson, "A Statistical Analysis of Surface Reverberation," *J. Acoust. Soc. Am.*, Vol. 74, pp. 249-255, 1983.

24. J. Wolfowitz, "The Minimum Distance Method," *Annals of Mathematical Statistics*, Vol. 28, pp. 75-88, 1957.

25. W.C. Parr and W.R. Schucany, "Minimum Distance and Robust Estimation," *J. Am. Stat. Assoc.*, Vol. 75, No. 371, pp. 616-624, 1980.

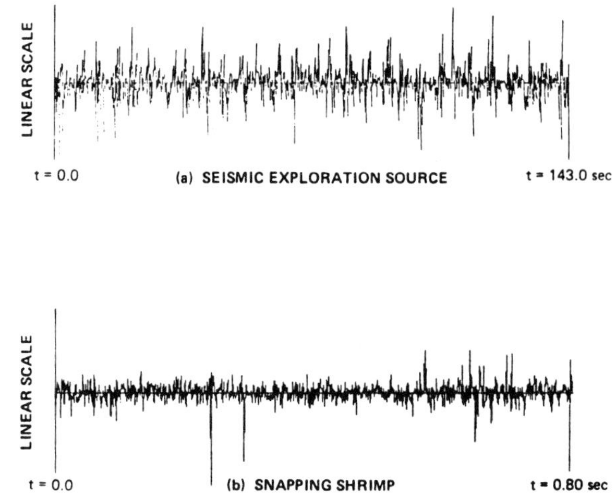

Figure 1 Time Series of Acoustic Data

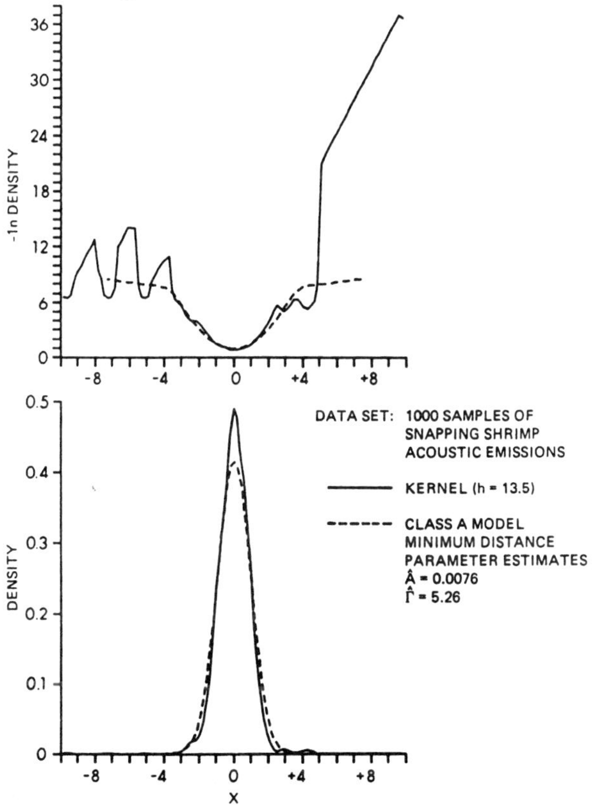

Figure 2 Comparison of Kernel and Class A Model Determined by Minimum Distance Criteria - Snapping Shrimp

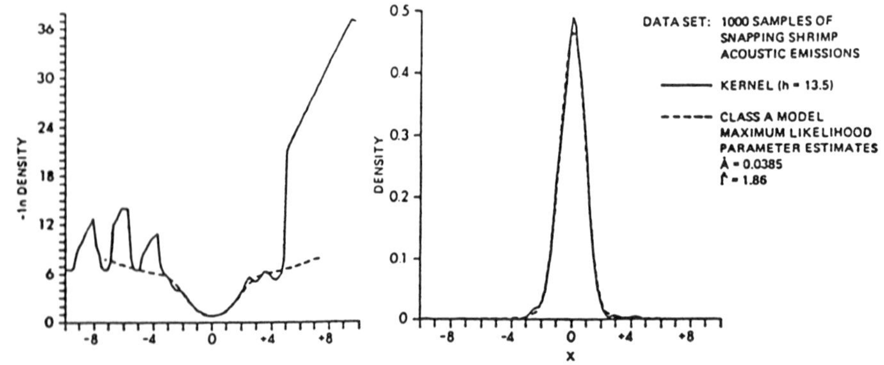

Figure 3 Comparison of Kernel and Class A Model Determined by Maximum Likelihood Criteria - Snapping Shrimp

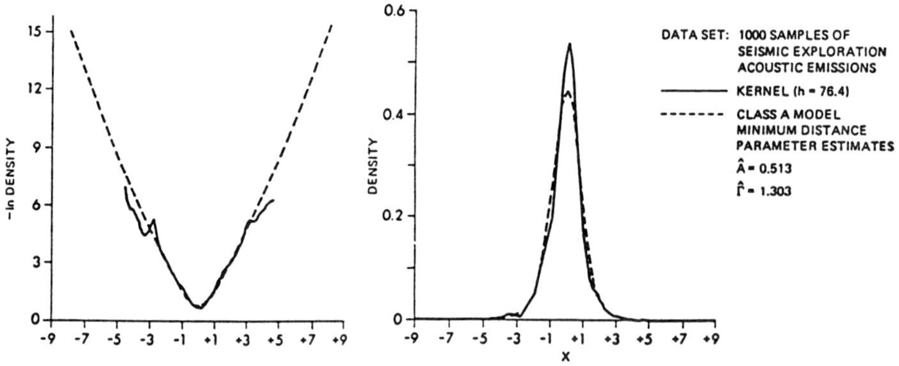

Figure 4 Comparison of Kernel and Class A Model Determined by Minimum Distance Criteria - Seismic Exploration Sources

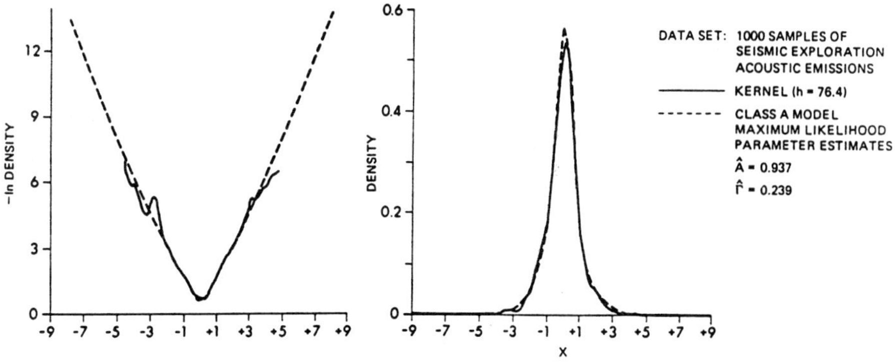

Figure 5 Comparison of Kernel and Class A Model Determined by Maximum Likelihood Criteria - Seismic Exploration Sources

TABLE I

Data	P-Values		
	Randomness	Homogeneity	
	Runs Test	KS-2 Test	Rank Sum Test
Seismic	.85	.62	.77
Snapping Shrimp	.74	.98	.57

TABLE II

Data	P-Values		
	Pearson Skew Test	Pearson Kurtosis Test	D'Agostino Test
Seismic	.47	0.00	0.00
Snapping Shrimp	.00	0.00	0.00

TABLE III

Data	Skew	Kurtosis
Seismic	-.06	5.4
Snapping Shrimp	-1.21	16.6

TABLE IV

N	PARAMETER	BIAS			VARIANCE			MEAN SQUARED ERROR		
		MOM	MD	ML	MOM	MD	ML	MOM	MD	ML
50	A	0.63	-0.11	0.06	2.48	2.2×10^{-3}	0.01	2.87	0.01	0.02
	Γ	2.3×10^{2}	0.12	-0.78	4.8×10^{5}	0.26	0.03	5.3×10^{5}	0.27	0.64
100	A	0.01	-0.13	-0.03	0.17	5.2×10^{-3}	4.6×10^{-3}	0.17	0.02	5.4×10^{-3}
	Γ	1.7×10^{3}	0.04	-0.80	2.4×10^{7}	0.22	8.2×10^{-3}	2.7×10^{7}	0.22	0.64
300	A	-0.20	-0.16	0.01	2.8×10^{-3}	1.3×10^{-3}	0.9×10^{-3}	0.04	0.03	1.0×10^{-3}
	Γ	60.2	-0.23	-0.88	1.3×10^{4}	0.01	1.0×10^{-3}	1.7×10^{4}	0.07	0.78

STATISTICAL CHARACTERISTICS OF OCEAN ACOUSTIC NOISE PROCESSES

Fredrick W. Machell, Clark S. Penrod, and Glen E. Ellis
Applied Research Laboratories
The University of Texas at Austin
Austin, Texas 78713-8029

ABSTRACT

A statistical analysis is given of ambient noise data from several ocean acoustic environments. Included in the analysis are statistical tests for homogeneity and randomness, statistical tests for normality, sample autocorrelation functions, and kernel density estimates of the instantaneous amplitude fluctuations. The test results indicate that a randomness hypothesis may be rejected when Nyquist rate sampling is employed. A randomization procedure is applied to the data in order to create ensembles which pass the tests for randomness and homogeneity. Analysis of these ensembles indicates that a stationary Gaussian assumption is not justified for some ocean environments. The largest deviations from normality occur in the tail regions of the density function and are often attributable to nonstationary characteristics of the data.

I. INTRODUCTION

In designing signal processing algorithms, knowledge of the statistical properties of the signal and noise processes is of paramount importance. In underwater acoustic signal processing these processes are usually assumed to be stationary and Gaussian, although some measurements indicate to the contrary (Ref. 1-5). This work examines the statistical characteristics of several ocean acoustic noise environments in order to test the validity of a stationary Gaussian assumption. Only a few papers have appeared in the literature dealing with measurement of the statistical properties of ambient noise in the ocean (Ref. 1-8). Thus, one purpose of this work is to expand the number of available measurements of noise statistics. The paper opens with a discussion of hypothesis testing techniques and data characterization tools used in this study. Next, a brief description of the data to be analyzed is given along with a discussion of procedures for generating ensembles which are both random and homogeneous. Finally, the statistical test results, autocorrelation functions, and kernel density estimates are presented for the five data sets.

II. HYPOTHESIS TESTING TECHNIQUES AND DATA CHARACTERIZATION TOOLS

All of the hypothesis tests considered here involve testing a single null hypothesis, H_0, against a single alternative hypothesis, H_1. For example, the test for randomness decides between the null hypothesis, H_0: the data are randomly generated, and the alternative hypothesis, H_1: the data are not randomly generated. Although there are many ways in which a collection of data can be nonrandom, no distinction is made between them in the test. In general, the tests employed here are such that the distribution of the test statistic under the null hypothesis is known or may be approximated. However, the power of each of the tests can not be determined since the distribution of the test statistic under the alternative hypothesis is unknown. Thus, when the alternative hypothesis is rejected, it really corresponds to not rejecting the null hypothesis on the basis of the sample ensemble tested (see Ref. 9).

In applying tests for normality, one should take care that the data are independent and identically distributed. Although independence is a difficult property to ascertain, a necessary condition for independence is that the data be uncorrelated or random. Hence, the runs up and down test for randomness (Ref. 10) is applied to the data to check for dependence in the data. This test counts the number of unbroken sequences of increasing or decreasing data points in the sample ensemble. Too few or too many runs indicate that the sample is not randomly generated and hence, an independence assumption is not justified. In order to check whether an ensemble is identically distributed, the Kolmogorov-Smirnov two-sample (KS2) test is employed. This test computes the largest absolute difference between the empirical distribution functions associated with the first half of the ensemble and the second half of the ensemble. Too large a difference in these two distributions indicates nonhomogeneities in the ensemble and hence, the identically distributed hypothesis may be rejected. The exact distribution of the test statistic for the KS2 test is known (Ref. 11), but the Smirnov approximation (Ref. 12) is adequate for large sample sizes and is used here. Note that the results of one of the tests may be influenced by results of the other test. For instance, a randomness hypothesis may be erroneously rejected due to nonhomogeneous data and vice-versa. Consequently, the test results should be interpreted carefully in order to avoid being led to the wrong conclusion.

The tests for normality used here are Pearson's test of skewness and Pearson's test of kurtosis (Ref. 13, 14). The skewness test statistic is simply the sample skewness:

$$\sqrt{\beta_1} = (m_3)/(m_2)^{1.5} \tag{1}$$

where m_k represents the kth sample moment about the mean. The kurtosis test statistic is the sample kurtosis given by:

$$\beta_2 = (m_4)/(m_2)^2 \tag{2}$$

The skewness test checks for nonsymmetric densities while the kurtosis test is sensitive to deviations from Gaussian in the tail regions of the density. The distribution of skewness under the null hypothesis (i.i.d. Gaussian data) converges quickly to a Gaussian distribution and hence, a normal approximation to this distribution is quite adequate for this work. Unfortunately, the distribution for kurtosis under the null hypothesis does not converge very fast and a normal approximation to this skew-leptokurtic distribution can result in significantly biased confidence intervals. The actual distribution has been approximated by a Pearson IV distribution yielding satisfactory results (Ref. 14). Confidence intervals based on the Pearson IV approximation will be used here.

When testing a set of ensembles for a particular property, it is desirable to determine if the entire collection of ensembles has that property. Several methods have been described in the literature for combining the results of a series of tests. Each of these methods operate on the fact that under the null hypothesis the p-value associated with the test statistic is uniformly distributed on the interval [0,1]. The p-value is simply the probability of exceeding the measured test statistic assuming that the ensemble belongs to the null hypothesis. Two methods are considered here for combining p-values. The first approach simply computes the sample mean of the p-values and compares the resulting statistic to the sampling distribution for the mean of independent uniform [0,1] random variables. The second method considers each p-value to be the result of an independent Bernoulli trial and computes the probability of obtaining k or greater number of trials below a prescribed threshold, where k is the number of observed p-values below the threshold. In all of the tests to follow a threshold of 0.1 is used. A discussion of these methods and additional references are provided by Wilson (Ref. 15).

A potential problem with combining test results arises if the test results are not independent. In this case, the combined p-values may be biased significantly. For example, if dependencies exist in the p-values the sample mean will probably have a larger variance associated with it thereby increasing the probability of obtaining a sample mean significantly different from the true mean value for the p-values. The runs up and down test was applied to the p-values showing significant dependence in some instances. To avoid the problems associated with dependent p-values, a combined p-value will be computed in the following way. The original p-values are randomized and then segmented into several groups. A combined p-value is computed for each of these groups and the average of these combined p-values is used as the overall combined p-value.

To complement the statistical tests, sample autocorrelation functions and kernel density estimates will be displayed. The sample autocorrelation functions given will consist of averages of estimates of the form:

$$R(n) = (1/N) \sum X_i * X_{i+n}; \; n = 0,1,...,N-1. \tag{3}$$

where $X_i (i=1,...,N)$ are the data samples. Sometimes a different factor $(1/(N-|n|))$ is given in the front of the sum of eq. (3), but as long as the time lag of interest (n) is small compared to the integration time (N) the difference is negligible. Prior to averaging, the sample autocorrelation functions will be normalized to the interval [-1,1] by dividing by the value at the zero lag, R(0), in eq. (3). The kernel estimates presented here are defined as:

$$f(x) = (1/Nh) \sum K[(x - X_i)/h] \tag{4}$$

where $K(\cdot)$ denotes a bounded continuous density function, $X_i (i=1,...,N)$ are the data points, and h is a positive parameter which controls the smoothness of the estimate (Ref. 16, 17). The kernel function used here is the Epanechnikov kernel (Ref. 18):

$$K(x) = 0.75(1 - x^2) \tag{5}$$

This kernel function is used because it gives the tightest bound on the rate of convergence of the L_1-error of the density estimate. For the smoothing factor, a value is selected which is optimal for Gaussian data in terms of minimum L_1-error (see Ref. 19, 20).

III. DESCRIPTION OF THE DATA AND ENSEMBLE FORMATION

In this study, data from five ocean ambient noise environments are analyzed. A brief description of the data is given followed by a discussion of formation of ensembles suitable for the analysis.

(1) QUIET DATA - collected in an extremely quiet region of the Pacific Ocean with no nearby shipping or other obvious nearfield noise sources.

(2) NOISY DATA - collected in the Indian Ocean, a region characterized by high shipping density, during a time period when no single dominant surface vessel could be identified. This region is one of the noisiest areas in the world.

(3) MERCHANT DATA - collected from the same area as (2) during a time period when a merchant vessel is in close proximity to the recording system. When the merchant approaches, the noise levels increase about 6-10 dB over the noisy levels of (2) throughout the frequency band analyzed (20-600 Hz).

(4) SEISMIC DATA - collected in the Gulf of Mexico during a time period of seismic exploration activities. The time series reflects the large noise bursts associated with the explorer's airguns. These bursts appear regularly throughout the data.

(5) ANTARCTIC DATA - collected in McMurdo Sound during the Antarctic spring (see Ref 21). Data is characterized by heavy biological activity (mainly the Weddell Sea seal and whales) and time series shows numerous transients.

As mentioned previously, in order to obtain meaningful results from the tests for normality, ensembles which are random and homogeneous are essential. These ensembles are constructed as follows. First, a time period consisting of 200,000 samples of data at the Nyquist rate is selected. Because the data were collected with different recording systems with different bandwidths, each data set was sampled at a different rate in the digitization process. Table 1 contains the sampling rates used. A 12-bit A/D converter was used for digitizing the data resulting in a dynamic range of about 66 dB. The effects of quantization are assumed negligible. Note that the Nyquist rate is somewhat arbitrary since ambient noise in the ocean has a energy spectrum which rolls off at about 6 dB/octave above 100-200 Hz (Ref. 22). In actuality, the sampling rates indicated in the table are probably twice the minimum rate at which the original waveform could be reconstructed from the samples with negligible error.

Table 1 - Sampling rates used in the digitization process

DATA SET	SAMPLE RATE (HERTZ)
QUIET DATA	1500
NOISY DATA	1250
MERCHANT DATA	1250
SEISMIC DATA	892
ANTARCTIC DATA	2400

As might be expected, with the sampling rates used the data show significant correlations. Furthermore, some of the data exhibits nonhomogeneities which could adversely affect the tests for normality. Consequently, a new collection of ensembles is formed for each data set by a randomization process as follows. First, the data are segmented into 200 blocks of length 1000. These 200 blocks are then shuffled using a random permutation vector generated from the IMSL statistical library. The resulting data set is regrouped into 64 blocks of length 3125 and then shuffled so that the mth group contains every 64th sample starting with sample m from the block permuted data. The 64 blocks are then regrouped into 40 blocks of length 5000. The data within these 40 blocks is randomly permuted using different permutation vectors from the IMSL package. For some of the data sets another shuffling was performed in order to obtain ensembles which were both homogeneous and random. Thus, two collections of ensembles are formed for each data set for the analysis and will be refered to throughout the paper as Nyquist and randomized data.

IV. STATISTICAL TEST RESULTS

In this section, the results of the statistical tests are presented for the five sets of Nyquist and randomized data. These results include sample moments as a function of time, combined p-values for the tests for homogeneity and randomness, combined p-values for the tests of skewness and kurtosis, sample autocorrelation functions, and kernel density estimates. Also included is a test for white noise described by Jenkins and Watts (Ref. 23).

Figures 1-5 contain the p-values which result from application of the KS2 test for homogeneity and the runs test for randomness for the five data sets. Segment lengths of 500, 1000, and 2000 samples were used for each of the tests but for brevity only the results for the 500 sample segment length are shown. The results for segment lengths of 1000 and 2000 show little difference from those for the 500 sample segment length. The biggest contrast is seen between the Nyquist and randomized data for the the runs test results. For the Nyquist data, the runs p-values are all nearly zero indicating that the data are not randomly generated. The results for the Antarctic data show some p-values significantly different from zero due to the duration of the impulsive noise bursts in these data. The bursts of impulsive noise for the Antarctic data are of much shorter duration than for the seismic data. These short bursts result in a larger variation of the number of runs for this data than for the other data sets. In addition, the different sampling rates of Tabel 1 seem to influence the results to some extent. A segment length of 500 corresponds to about 208 msec for the Antarctic data and about 561 msec for the seismic data. With larger segment lengths, a larger percentage of p-values are zero for the Antarctic data.

Table 2 contains the combined p-values for the KS2 and runs tests for the Nyquist data. The deep ocean quiet data is seen to be the only data set which passes the homogeneity test with certainty, while the noisy and merchant data sets marginally pass the test. The seismic and the Antarctic data sets both fail the homogeneity test with combined p-values of zero primarily due to the nonstationarities associated with these data. All of the data sets fail the runs test for randomness with combined p-values of zero due to the correlation between the samples. The sample autocorrelation functions and white noise test given later give a clear indication of this correlation. Table 3 contains the combined p-values for the KS2 and runs tests for the randomized data. All of the data sets are seen to pass both tests with certainty after randomization. Table 4 shows the sample mean and variance of the number of runs for the randomized and Nyquist data. In addition, the theoretical value of the mean and variance of the runs statistic under the null hypothesis is given. As can be seen, the Nyquist data fails the test due to consistently too few runs. The variance of the number of runs is excessively large for the seismic and Antarctic data due to the impulsive characteristics of these data. For the randomized data, the runs statistic is seen to have a sample mean and variance consistent with the null hypothesis.

Table 2 - Combined p-values for KS2 and runs tests for Nyquist data

DATA SET	KS2 TEST MEAN	KS2 TEST BINOMIAL	RUNS TEST MEAN	RUNS TEST BINOMIAL
QUIET	0.91	0.99	0.00	0.00
NOISY	0.03	0.17	0.00	0.00
MERCHANT	0.12	0.27	0.00	0.00
SEISMIC	0.00	0.00	0.00	0.00
ANTARCTIC	0.00	0.00	0.00	0.00

Table 3 - Combined p-values for KS2 and runs tests for randomized data

DATA SET	KS2 TEST MEAN	KS2 TEST BINOMIAL	RUNS TEST MEAN	RUNS TEST BINOMIAL
QUIET	0.32	0.52	0.51	0.28
NOISY	0.68	0.63	0.55	0.32
MERCHANT	0.34	0.37	0.84	0.95
SEISMIC	0.29	0.53	0.34	0.35
ANTARCTIC	0.73	0.76	0.30	0.45

Table 4 - Mean and variance of runs test statistic for segment length of 500

DATA SET	NYQUIST MEAN	NYQUIST VARIANCE	RANDOMIZED MEAN	RANDOMIZED VARIANCE
THEORY	--	--	333.0	88.6
QUIET	270	57.4	332.5	94.0
NOISY	239	169.	331.7	93.5
MERCHANT	212	412.	333.1	85.7
SEISMIC	144	76.	332.0	95.5
ANTARCTIC	293	3488.	330.9	86.3

Figures 6-10 show sample moments as a function of time for the Nyquist and randomized data. These moments are based on contiguous blocks of length 500. Several features are worth mention. First, the variations in the sample moments are more uniform after randomization. The variance of the sample mean increases after randomization of the quiet, noisy, and merchant data sets. Interestingly, the variance of the sample mean is consistent with a white noise assumption after randomization, while it is significantly smaller than expected with white noise prior to randomization of these data. In addition, the moments of the seismic and the Antarctic data clearly show the nonstationary characteristics associated with these data. Furthermore, the large negative skew values of the Antarctic data are in direct correspondence with the large kurtosis values. The Antarctic data also shows a large number of kurtosis values significantly less than the Gaussian value of 3 indicating the presence of strong tonals in the data. These tonals are probably associated with the biological activity in the area. The kurtosis values for the merchant data are consistently less than the Gaussian value of 3. These small kurtosis values may be a result of the tonals radiated by the merchant.

Table 5 summarizes the sample moments of the Nyquist data averaged over the 400 blocks of length 500. Table 6 contains the corresponding values for the randomized data. The randomization process has little effect on the averaged sample moments with the exception of the skew and kurtosis values for the seismic and Antarctic data sets. For these data, the impulsive bursts are the source of this difference. Randomization tends to spread out the large samples associated with the bursts uniformly across the data set so that a larger percentage of data segments contain outliers. As a result, the randomization procedure increases the averaged skew and kurtosis values while reducing the variances of the sample skew and kurtosis for these data.

Table 5 - Sample moments of Nyquist data averaged over 400 blocks of length 500

DATA SET	MEAN	VARIANCE	SKEWNESS	KURTOSIS
QUIET	2.64	4.33×10^4	0.013	2.94
NOISY	-0.717	7.05×10^4	-0.018	3.05
MERCHANT	-0.265	2.83×10^5	0.021	2.30
SEISMIC	-2.90	5.88×10^4	-0.029	3.31
ANTARCTIC	0.778	8.55×10^3	-0.106	3.67

Table 6 - Sample moments of randomized data averaged over 400 blocks of length 500

DATA SET	MEAN	VARIANCE	SKEWNESS	KURTOSIS
QUIET	2.64	4.32×10^4	0.014	2.97
NOISY	-0.717	7.04×10^4	-0.021	3.10
MERCHANT	-0.265	2.82×10^5	0.020	2.31
SEISMIC	-2.90	5.89×10^4	-0.108	5.13
ANTARCTIC	0.778	8.56×10^3	-0.079	4.83

Table 7 contains the combined p-values for Pearson's test of skewness for the Nyquist and randomized data. Both the merchant and the quiet data pass the combined test with ease while the noisy data passes the test marginally. In contrast, the seismic and the Antarctic data give combined p-values of zero due to the large number of skewed ensembles (see the sample skewness of Fig. 9-10). Note the difference in combined p-values when Nyquist sampling and random sampling are employed for some of the data sets. This difference may be due to the dependence present when Nyquist sampling is used.

Table 7 - Combined p-values for Pearson's test of skewness

	NYQUIST		RANDOMIZED	
DATA SET	MEAN	BINOMIAL	MEAN	BINOMIAL
QUIET	0.19	0.15	0.49	0.56
NOISY	0.07	0.08	0.15	0.17
MERCHANT	1.00	1.00	1.00	1.00
SEISMIC	0.00	0.00	0.00	0.00
ANTARCTIC	0.00	0.00	0.00	0.00

Table 8 shows the combined p-values for Pearson's test of kurtosis for the Nyquist and randomized data. The quiet data is the only data set which passes the test and the noisy data is the only other data set with a nonzero combined p-value for kurtosis. Only the results for the binomial method are shown because of the difficulty of computing the Pearson IV approximation to the sampling distribution for kurtosis. Thresholds of 0.05 and 0.1 are shown for this test. The values shown do not contain as much information as the p-values based on the mean since they are based only on those p-values which fall below 0.05 or 0.1. Nonetheless, on the basis of these results a Gaussian assumption may be conclusively rejected for the merchant, seismic and Antarctic data. The rejection for the noisy data is not as conclusive.

Table 8 - Combined p-values for Pearson's test of skewness

	NYQUIST		RANDOMIZED	
DATA SET	p=0.05	p=0.1	p=0.05	p=0.1
QUIET	0.80	0.64	0.57	0.59
NOISY	0.00	0.02	0.07	0.01
MERCHANT	0.00	0.00	0.00	0.00
SEISMIC	0.00	0.00	0.00	0.00
ANTARCTIC	0.00	0.00	0.00	0.00

Figure 11 displays the first 200 time lags of averaged sample autocorrelation functions for the Nyquist and randomized data. The curves shown are based on an average of 100 autocorrelation functions with 2000 samples per autocorrelation function. The curves for the Nyquist data show significant correlations with features similar to the exponential-cosine correlation functions of Bendat (Ref. 24). The autocorrelation function for the Antarctic data reveals the strong tonals at 10 and 60 Hz. The 60 Hz tone can be traced to electronic system noise. In contrast, the sample autocorrelation functions are nearly identically zero for the randomized data showing the lack of correlation of this data.

The correlation functions of Figure 11 provide additional information about the data. Jenkins and Watts (Ref. 23) describe a test for white noise based on the sample autocorrelation function. The test uses the fact that the sample autocorrelation function is asymptotically normal with mean zero and a variance given by:

$$Var\{R(n)\} = (N - |n|)/(NAVG \cdot N^2) \tag{6}$$

where n is the time lag of interest, N is the integration time for an individual correlation function, and NAVG is the number of correlation functions averaged. Eq. (6) is based on the assumption that the individual correlations averaged are independent. Using the variance of eq. (6), 95% confidence intervals can be constructed for the autocorrelation lags. Using a normal approximation, the confidence intervals are $\pm\, 1.96 \cdot \sqrt{Var\{R(n)\}}$. Table 7 shows the percentage of the first 1000 lags falling outside of the 95% intervals for the Nyquist and randomized data.

Table 9 - Percentage of autocorrelation lags failing test for white noise at 5% significance level

DATA SET	NYQUIST	RANDOMIZED
QUIET	40.4	6.0
NOISY	64.9	5.7
MERCHANT	55.5	6.3
SEISMIC	89.7	5.2
ANTARCTIC	97.6	4.7

Figures 12-16 show kernel density estimates for the five data sets. Part (a) of the figures show 40 kernel estimates for the Nyquist data with each estimate based on 5000 samples of data. These estimates show the fluctuations of the instantaneous amplitude statistics of the data. Part (b) shows the corresponding estimates for the randomized data. The estimates are very similar for the Nyquist data in part (a) and for the randomized data in part (b) for all but the seismic and the Antarctic data. For these data, the nonstationary characteristics are readily apparent in the density estimates.

The average of the estimates in (a) are shown on a linear scale in part (c) and on a log scale in part (d). The linear scale enhances features of the density around the mode while the logarithmic scale shows the structure of the tails of the density. The averaged (nonparametric) kernel density estimates appear as solid curves and generalized Gaussian densities (parametric estimates) are represented by symbols in the figures. The generalized Gaussian family of density functions (Ref. 25) is given by

$$f(x) = c\,\eta(\sigma,c)/(2 \cdot \Gamma(1/c)) \cdot \exp\{1[\eta(\sigma,c) \cdot |x|]^c\} \tag{7}$$

where

$$\eta(\sigma,c) = (1/\sigma) \cdot [\Gamma(3/c)/\Gamma(1/c)]^{1/2} \tag{8}$$

$\Gamma(\cdot)$ is the gamma function, σ^2 is the variance of the generalized Gaussian density, and c is the decay rate parameter. This family of densities allows for different exponential rates of decay in the tails of the density and includes the Gaussian density (c=2) and the Laplace density (c=1). A Gaussian density with the same mean and variance as the sample mean and sample variance of the data is fitted to the density estimates for the quiet and noisy data sets. The other three data sets are fit with a generalized Gaussian density with the decay rate parameter chosen by selecting the density in the family with kurtosis closest to the kurtosis values of the randomized data (see Table 5). The exponential decay rates used were c=3.36 for the merchant data, c=1.18 for the seismic data, and c=1.125 for the Antarctic data. These generalized Gaussian densities give good agreement with the density estimates particularly in the tail regions.

V. SUMMARY

In summary, a statistical analysis of recorded ambient noise data from five ocean acoustic environments has been given. These noise data included data from a very quiet region of the world, data from a noisy region of the world, two examples of data corrupted by man-made noise, and data corrupted by biological noise in the Antarctic. The analysis included statistical tests for homogeneity and randomness, statistical tests for normality, sample moments up to fourth order, estimates of the autocorrelation function, and density estimates of the instantaneous amplitude fluctuations.

The data from the quiet region displayed characteristics which were the closest to the usual assumption of stationary Gaussian statistics. These data showed the least statistical fluctuations of the data sets analyzed. In addition, this was the only data set to convincingly pass the statistical tests for normality, a result which might be anticipated. In the absence of any dominant near-field noise sources where the primary contributions to the noise field are from local wind and distant shipping conditions the Central Limit Theorem (CLT) should drive any non-normalities toward Gaussian.

The data from the noisy region displayed characteristics which were fairly close to Gaussian with more fluctuations than the data from the quiet region. As with the quiet data set, there were no distinguishable near-field sources which dominated the noise field. The tests for normality were inconclusive with the data showing some tendency toward heavier tails than Gaussian. Density estimates confirmed this tendency toward a heavier tailed density. The observed deviations from Gaussian are probably attributable to the heavy shipping traffic of the area as the number of significant noise sources contributing to the local noise field were insufficient to obtain convergence of the CLT in the tails of the density.

The merchant dominated data showed a significant departure from the usual Gaussian assumption. These data were characterized by small values of kurtosis indicating smaller tails than Gaussian. Density estimates constructed from the data were in agreement with this observation. The source of this nonnormality is probably the tonals radiated by the merchant which are generated by onboard machinery and propeller cavitation.

The data corrupted by the noise of the seismic exploration vessel showed significant departure from normality and stationarity. These data were characterized by bursts of impulsive noise appearing at regular intervals, resulting in larger values of kurtosis than normal. Some of the ensembles also showed large skew values with the onset of the impulsive bursts. Density estimates confirmed the heavy tailed density associated with the data.

The Antarctic data showed the largest deviations from stationarity and Gaussianity primarily due to biological activity. These nonstationary noise bursts occured with no regularity and were much shorter in duration than those associated with the seismic data. Very large values of skew and kurtosis could be associated with the ensembles containing the impulsive noise bursts. Density estimates revealed large deviations from normality particularly in the tail regions.

The nonstationary and non-Gaussian characteristics of some of the data analyzed could adversely affect the performance of systems designed under the usual stationary, Gaussian noise assumption. With large deviations in the tails of the noise density, the probability of error in such a system is almost certainly going to differ from theoretical predictions. A related study (Ref. 26) describes the kind of performance degradation that may result from operation in such environments. Hopefully, these degradations may be reduced to some extent through the use of robust signal processing algorithms.

ACKNOWLEDGEMENT

This work was supported by the Statistics and Probability Program of the Office of Naval Research. The authors are thankful for the helpful discussions with Dr. Gary Wilson and Dr. Patrick Brockett, and for the help of Dr. Alick Kibblewhite, visiting scientist from the University of Auckland, New Zealand, in identifying the biological sources in the Antarctic data.

REFERENCES

1. T. Arase and E. M. Arase, "Deep-Sea Ambient Noise Statistics," *J. Acoust. Soc. Am.,* Vol. 44, pp. 1679-1684, (1968).

2. A. R. Milne and J. H. Ganton, "Ambient Noise Under Arctic-Sea Ice," *J. Acoust. Soc. Am.,* Vol. 36, pp. 855-863, (1964).

3. R. Dwyer, "FRAM II Single Channel Ambient Noise Statistics," NUSC Technical Document 6583, (1981).

4. J. G. Veitch and A. R. Wilks, "A Characterization of Arctic Undersea Noise," Dept. of Statistics Tech. Rep. 12, Princeton Univ., (1983).

5. F. W. Machell and C. S. Penrod, "Probability Density Functions of Ocean Acoustic Noise Processes," in *Statistical Signal Processing,* edited by E. J. Wegman and J. G. Smith, Marcel Dekker, New York, (1984).

6. W. J. Jobst and S. L. Adams, "Statistical Analysis of Ambient Noise," *J. Acoust. Soc. Am.,* Vol. 62, pp. 63-71, (1977).

7. A. H. Green, "A Study of the Bivariate Distribution of Ocean Noise," Bell Telephone Lab., Inc., Tech. Rep. No. 10, (1962).

8. M. A. Calderon, "Probability Density Analysis of Ocean and Ship Noise," U. S. Navy Electron. Lab., San Diego, CA, Rep. 1248, (1964).

9. G. R. Wilson, "Covariance Functions and Related Statistical Properties of Acoustic Backscattering from a Randomly Rough Air-Water Interface," Applied Research Laboratories Tech. Rep No. ARL-TR-81-23, (1981), also as Ph. D. Dissertation, Univ. of Texas at Austin, (1981).

10. J. V. Bradley, Distribution–Free Statistical Tests, Prentice, Englewood Cliffs, NJ, (1968).

11. P. J. Kim and R. I. Jennrich, "Tables of the Exact Sampling Distribution of the Two-Sample Kolmogorov-Smirnov Criterion, D_{nm}, $n \leq m$," in *Selected Tables in Mathematical Statistics,* edited by the Institute of Mathematical Statistics, American Mathematical Society, Providence RI, vol. 1, 79-90, (1970).

12. P. J. Kim, "On the Exact and Approximate Sampling Distribution of the Two-Sample Kolmogorov-Smirnov Criterion, $D_{nm}, m \leq n$," *J. Am. Stat. Assoc.,* Vol. 64, pp. 1625-1637, (1969).

13. E. S. Pearson, "A Further Development of Tests for Normality," *Biometrika 22,* pp. 239-249, (1930).

14. E. S. Pearson, "Note on Tests for Normality," *Biometrika 22,* pp. 423-424, (1930).

15. G. R. Wilson, "A Statistical Analysis of Surface Reverberation," *J. Acoust. Soc. Am.,* Vol. 74, pp. 249-255, (1983).

16. E. Parzen, "On Estimation of a Probability Density Function and the Mode," *Annals of Math. Stat. 33,* pp. 1065-1076, (1962).

17. M. Rosenblatt, "Remarks on Some Nonparametric Estimates of a Density Function," *Annals of Math. Stat. 27,* pp. 832-835, (1956).

18. V. A. Epanechnikov, "Nonparametric Estimates of a Mulitvariate Probability Density Function," *Theory of Probability and its Applications 14,* pp. 153-158, (1969).

19. L. Devroye, F. Machell, and C. Penrod, "The Transformed Kernel Estimate," submitted to *J. Am. Stat. Assoc.,* (1984).

20. L. Devroye and L. Gyõrfi, Nonparametric Density Estimation: The L_1 View, to be published in the Wiley Series in Probability and Mathematical Statistics.

21. A. C. Kibblewhite and D. A. Jones, "Ambient Noise Under Antarctic Sea Ice," *J. Acoust. Soc. Am. 59,* pp. 790-798, (1976).

22. R. J. Urick, Principles of Underwater Sound, 2nd ed., McGraw-Hill, New York, (1967).

23. G. M. Jenkins and D. G. Watts, Spectral Analysis and its Applications, Holden-Day, San Francisco, CA, (1968).

24. J. S. Bendat, Principles and Applications of Random Noise Theory, Wiley, London, (1958).

25. J. H. Miller and J. B. Thomas, "Detectors for Discrete-Time Signals in Non-Gaussian Noise," *IEEE Trans. on Information Theory,* Vol. IT-18, pp. 241-250, (1972).

26. F. W. Machell and C. S. Penrod, "Energy Detection in the Ocean Acoustic Environment," paper in this volume.

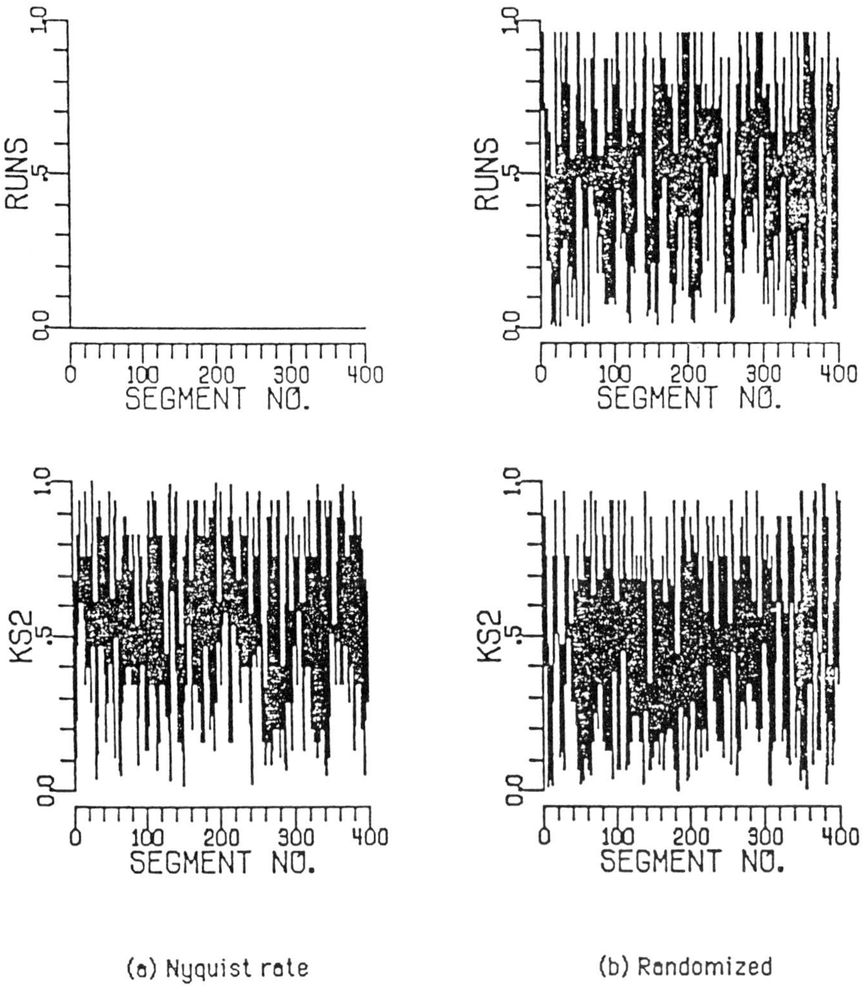

FIGURE 1

P-VALUES FOR QUIET DATA

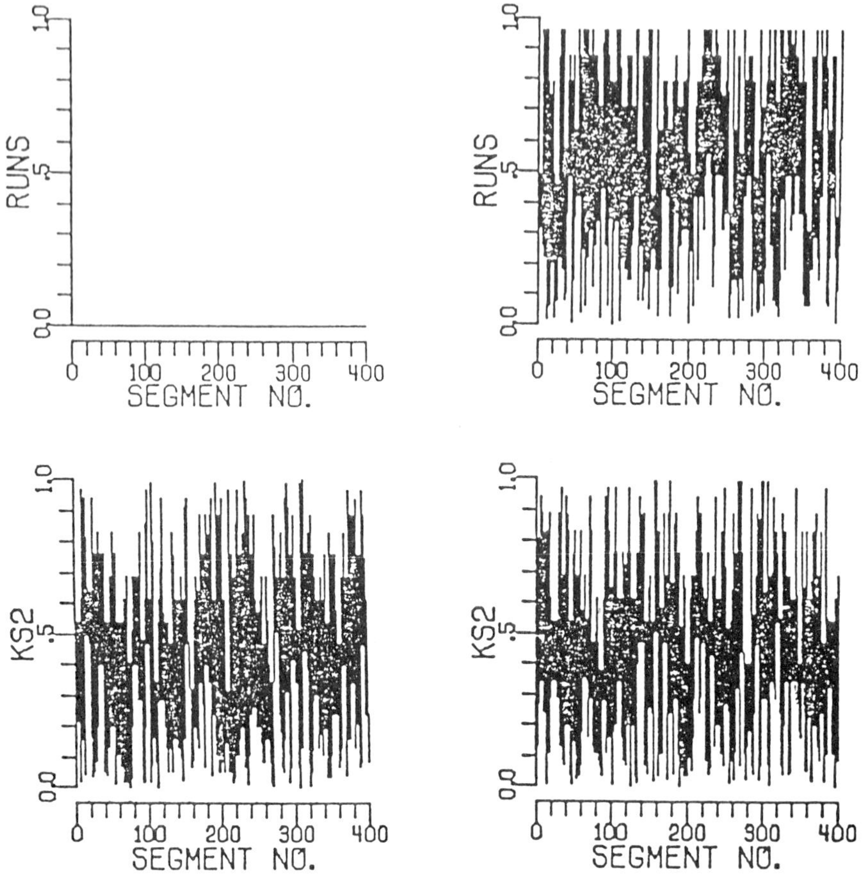

(a) Nyquist rate (b) Randomized

FIGURE 2

P-VALUES FOR NOISY DATA

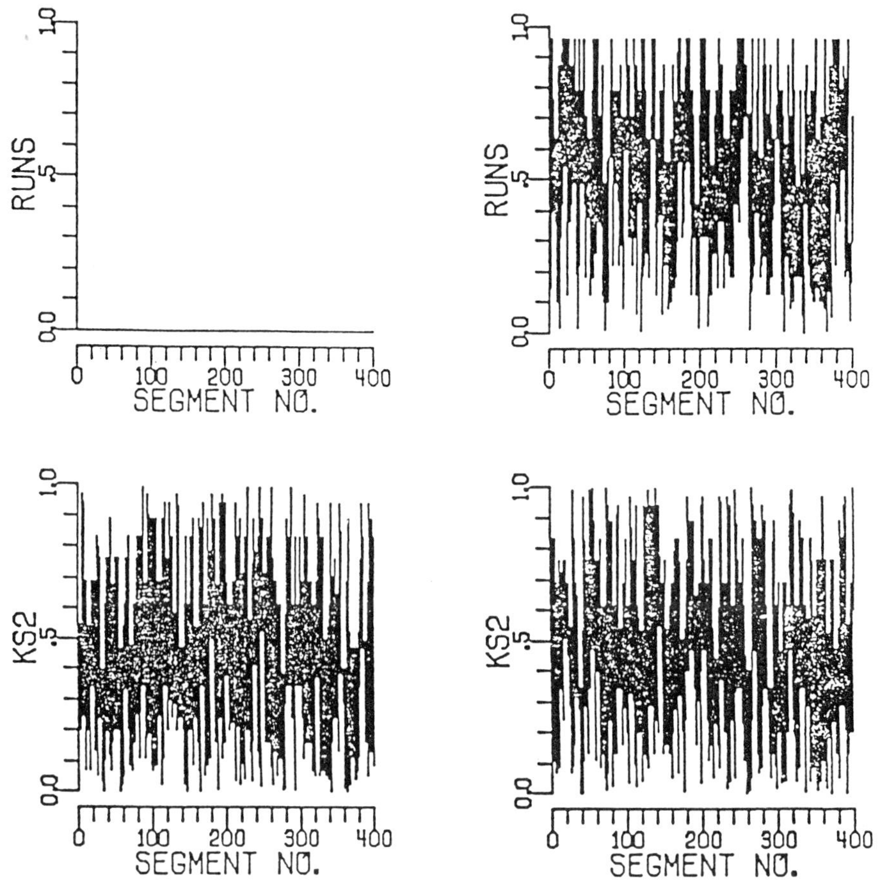

(a) Nyquist rate

(b) Randomized

FIGURE 3

P-VALUES FOR MERCHANT DATA

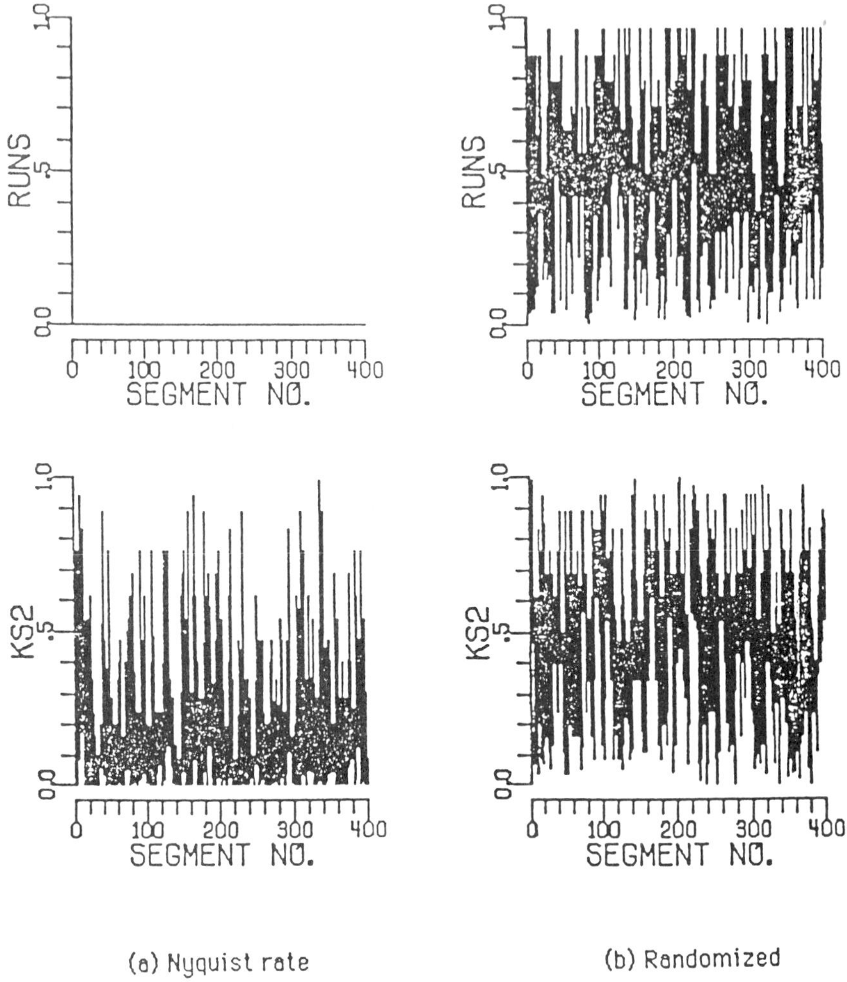

FIGURE 4

P-VALUES FOR SEISMIC DATA

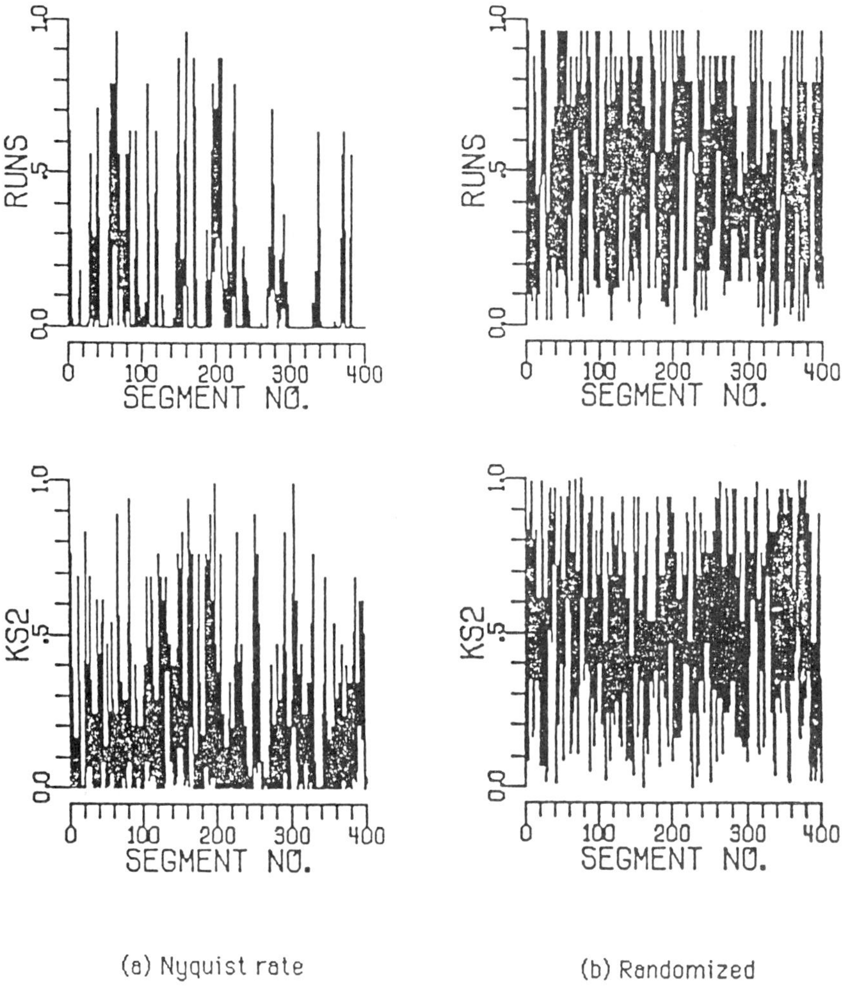

FIGURE 5

P-VALUES FOR ANTARCTIC DATA

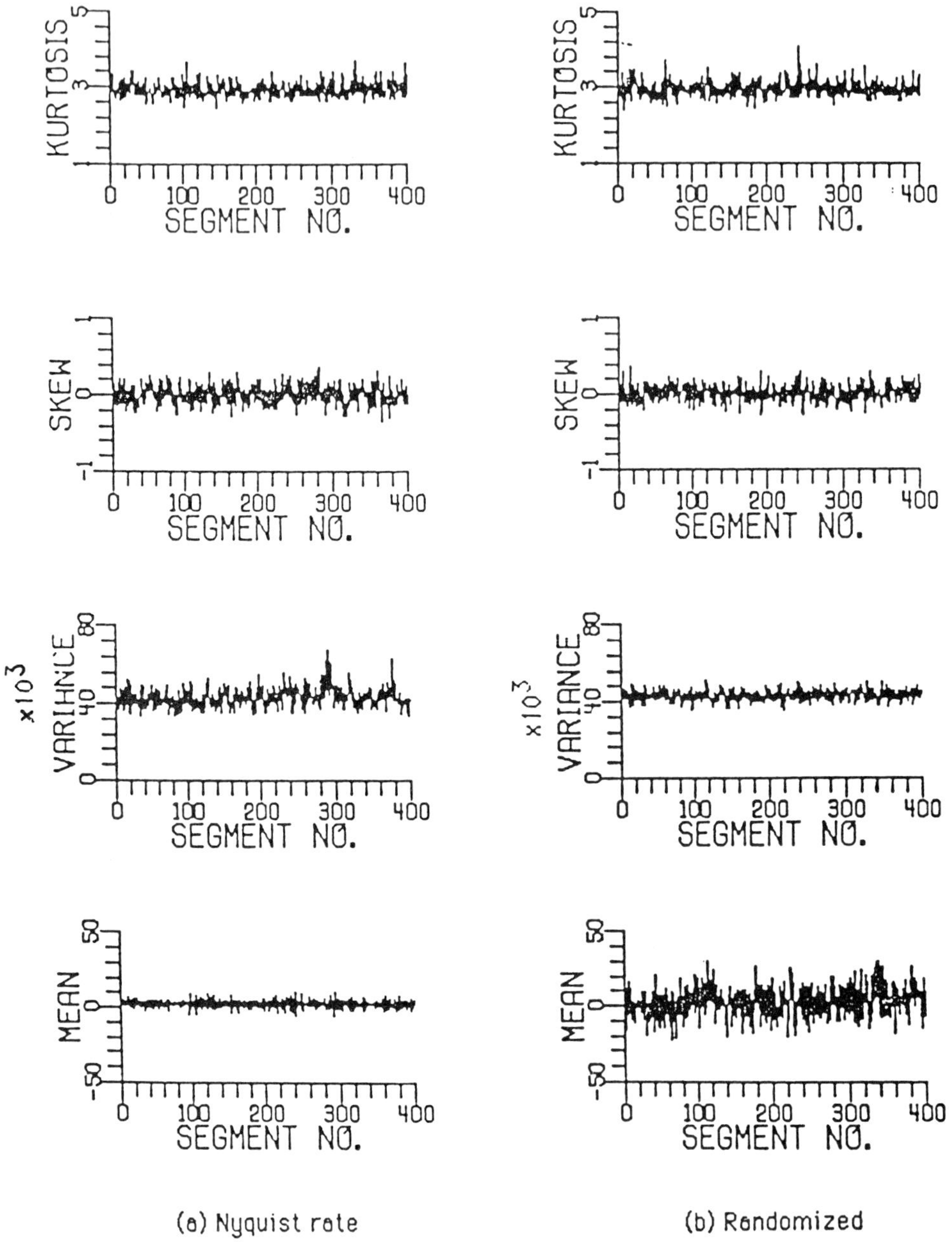

FIGURE 6

SAMPLE MOMENTS FOR QUIET DATA

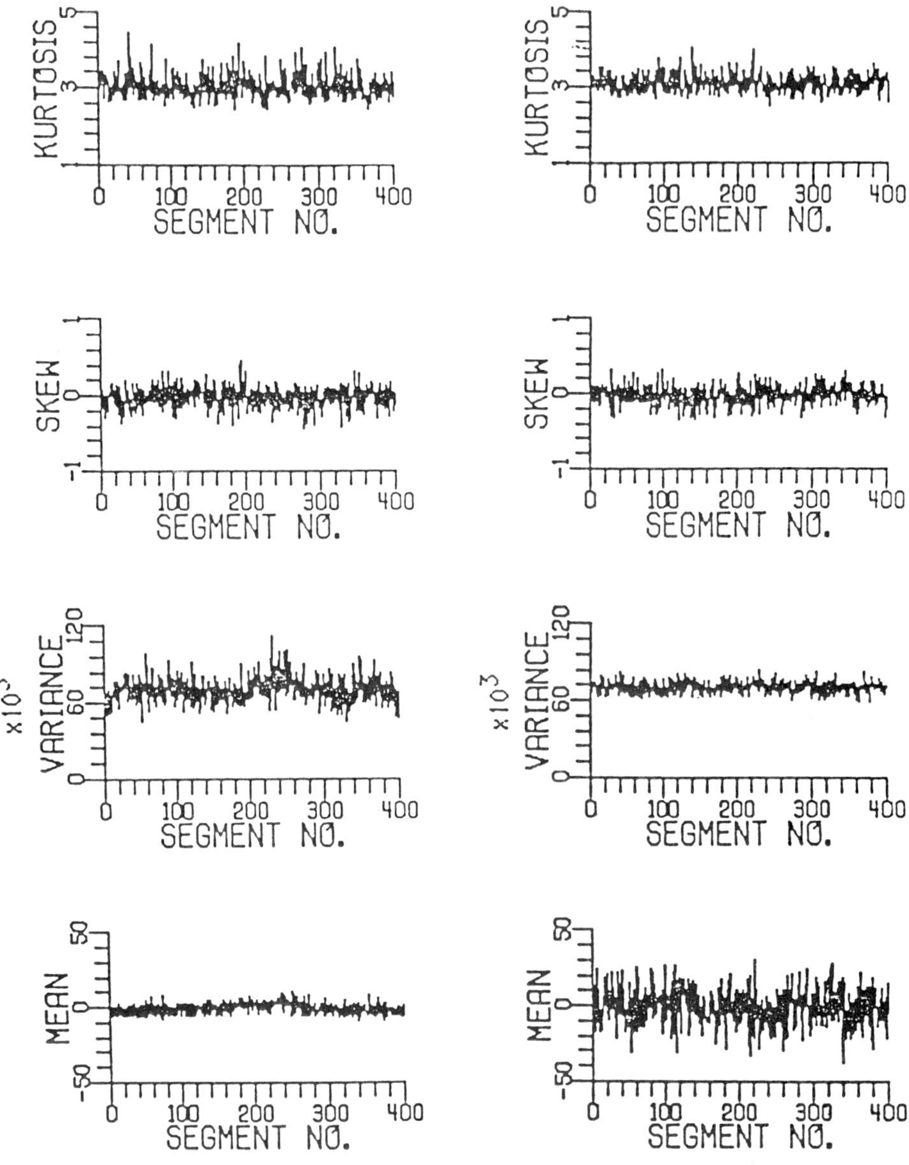

(a) Nyquist rate (b) Randomized

FIGURE 7

SAMPLE MOMENTS FOR NOISY DATA

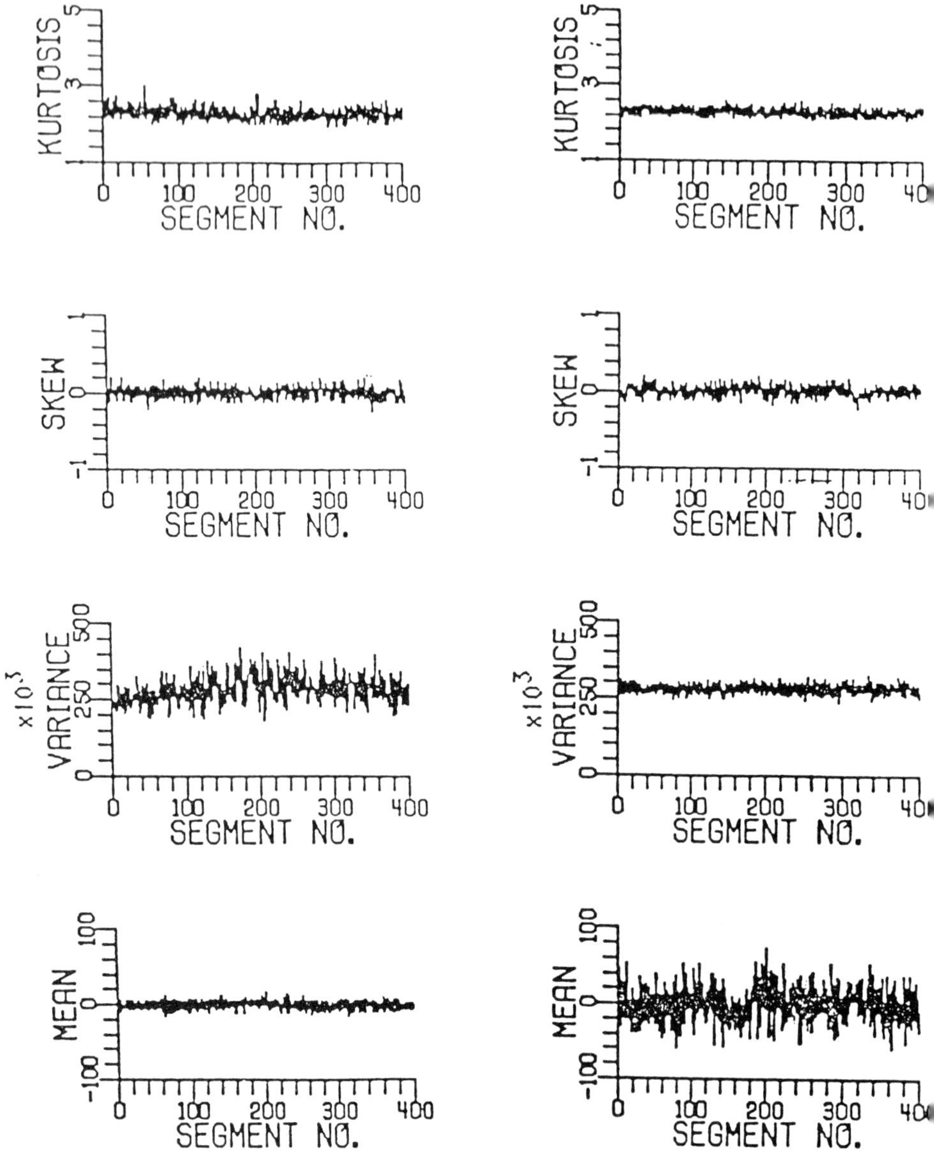

(a) Nyquist rate

(b) Randomized

FIGURE 8

SAMPLE MOMENTS FOR MERCHANT DATA

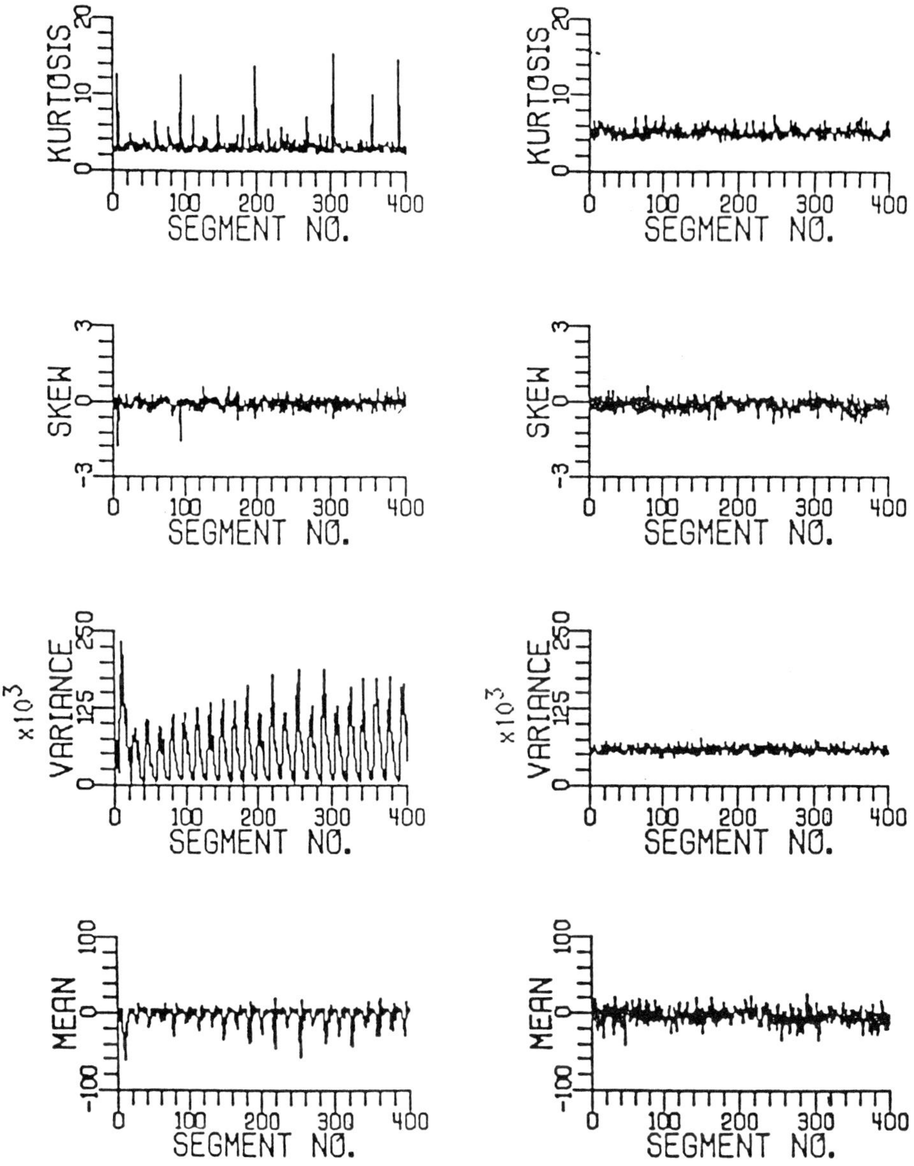

(a) Nyquist rate

(b) Randomized

FIGURE 9

SAMPLE MOMENTS FOR SEISMIC DATA

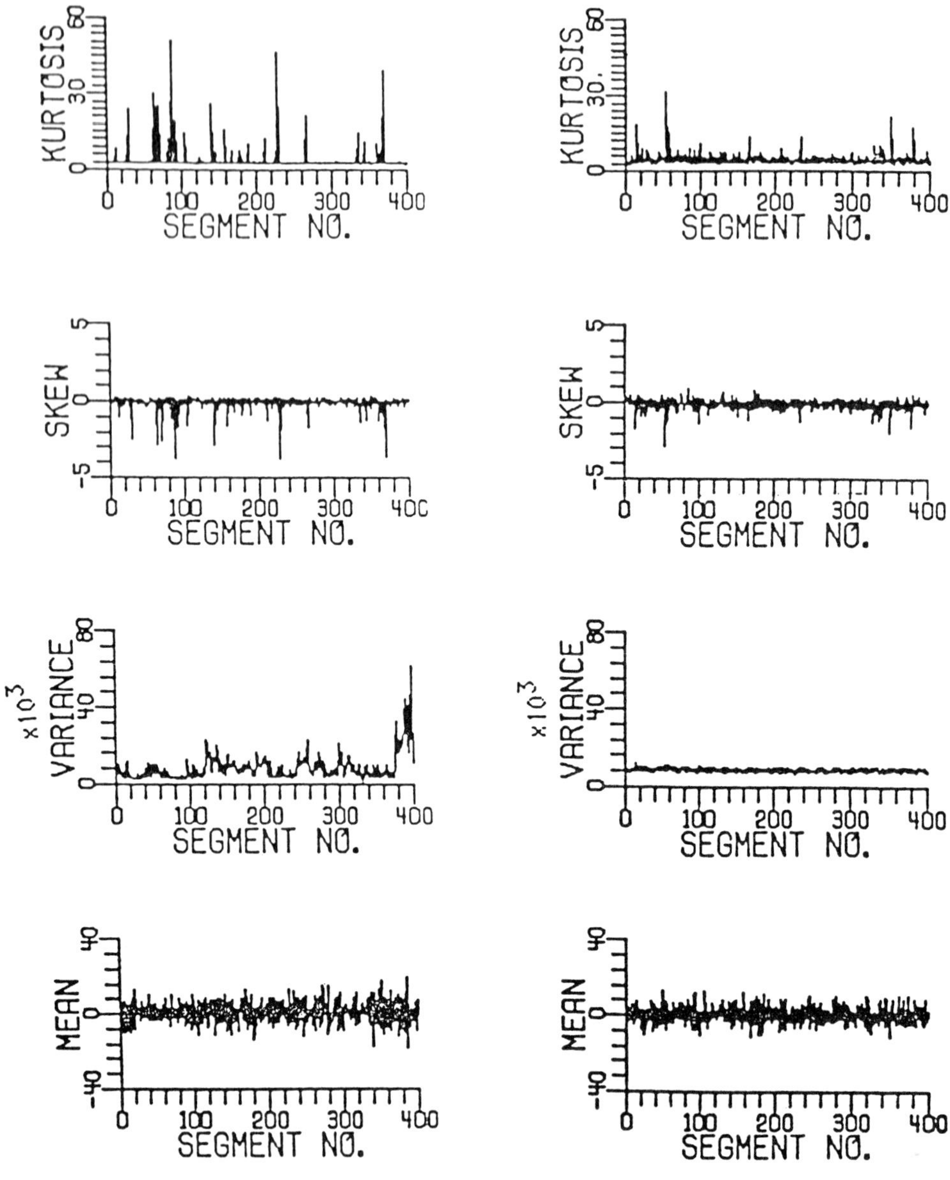

(a) Nyquist rate

(b) Randomized

FIGURE 10

SAMPLE MOMENTS FOR ANTARCTIC DATA

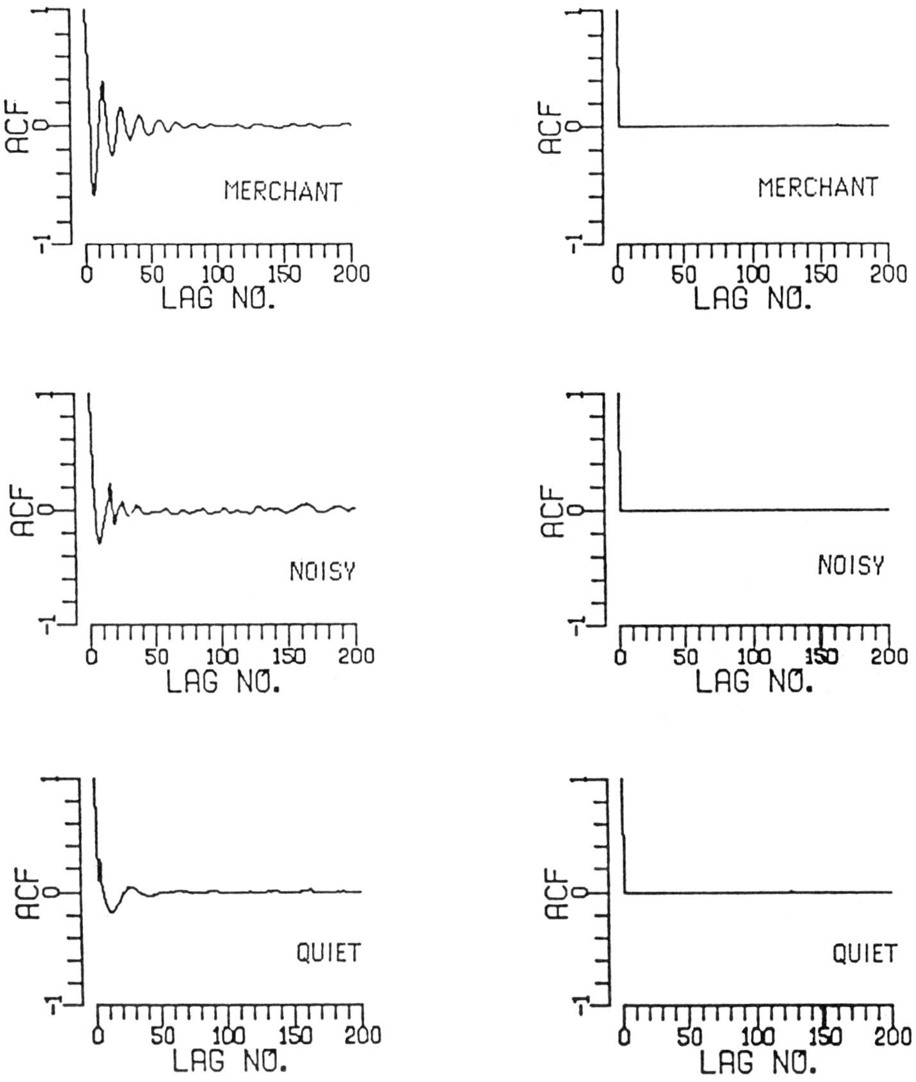

(a) Nyquist rate

(b) Randomized

FIGURE 11

SAMPLE AUTOCORRELATION FUNCTIONS

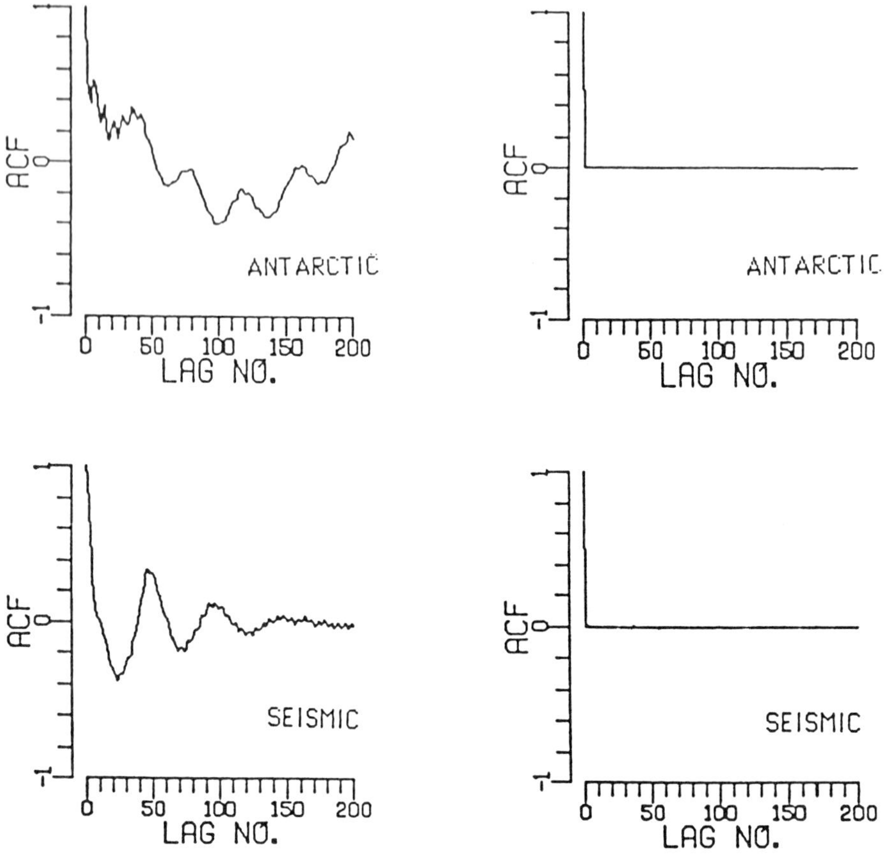

(a) Nyquist rate

(b) Randomized

FIGURE 11 (cont)

SAMPLE AUTOCORRELATION FUNCTIONS

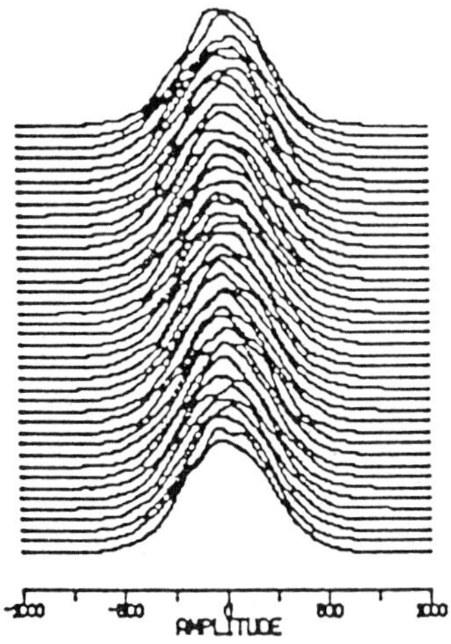

(a) Nyquist rate

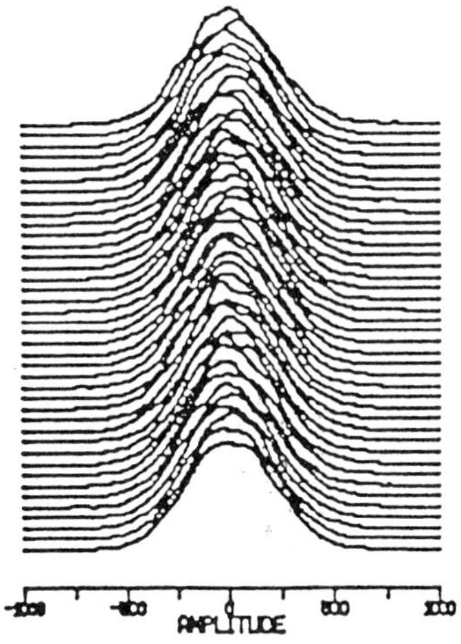

(b) Randomized

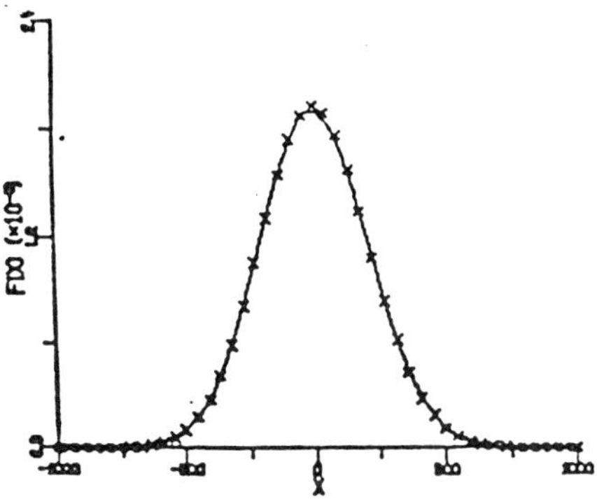

(c) Average on linear scale

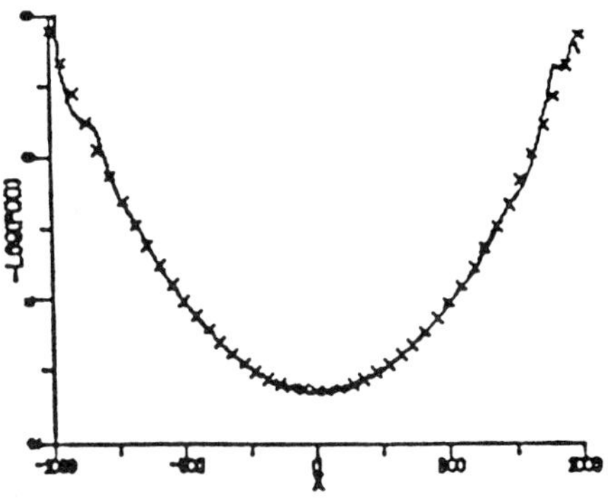

(d) Average on logarithmic scale

FIGURE 12

KERNEL DENSITY ESTIMATES FOR QUIET DATA

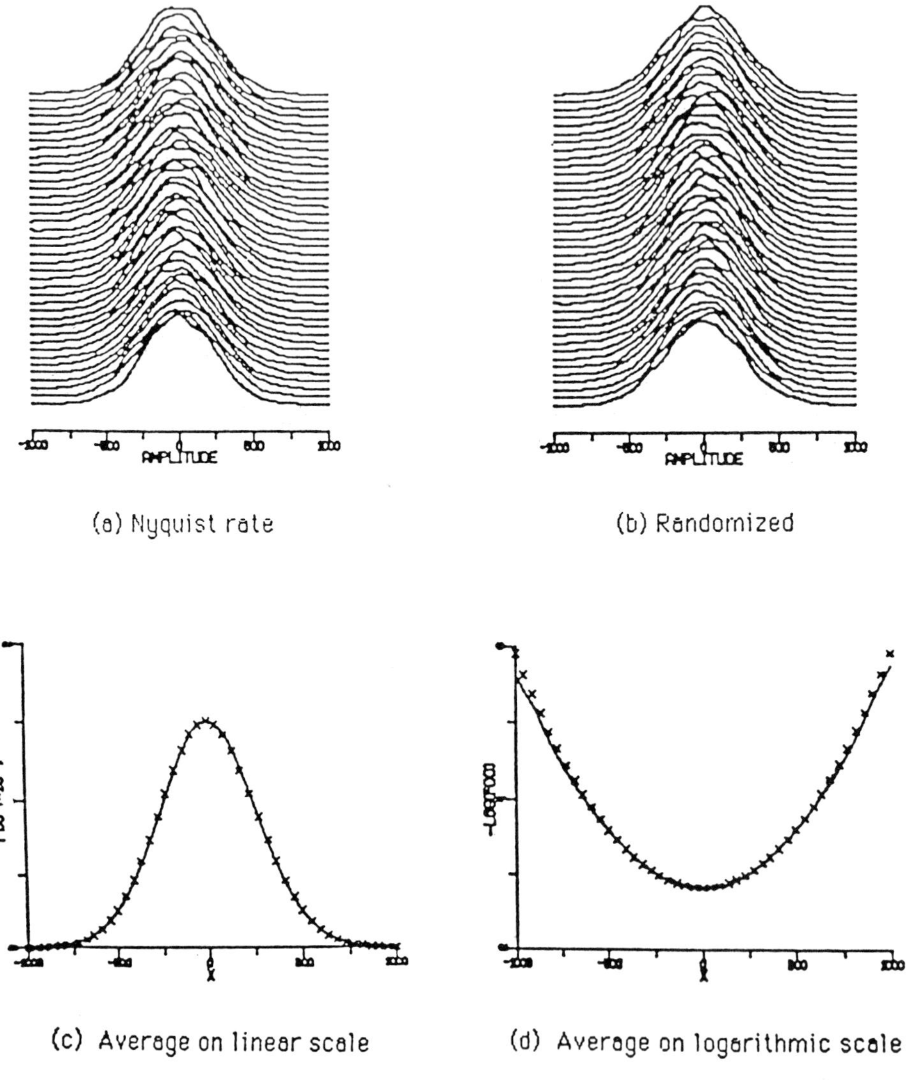

(a) Nyquist rate

(b) Randomized

(c) Average on linear scale

(d) Average on logarithmic scale

FIGURE 13

KERNEL DENSITY ESTIMATES FOR NOISY DATA

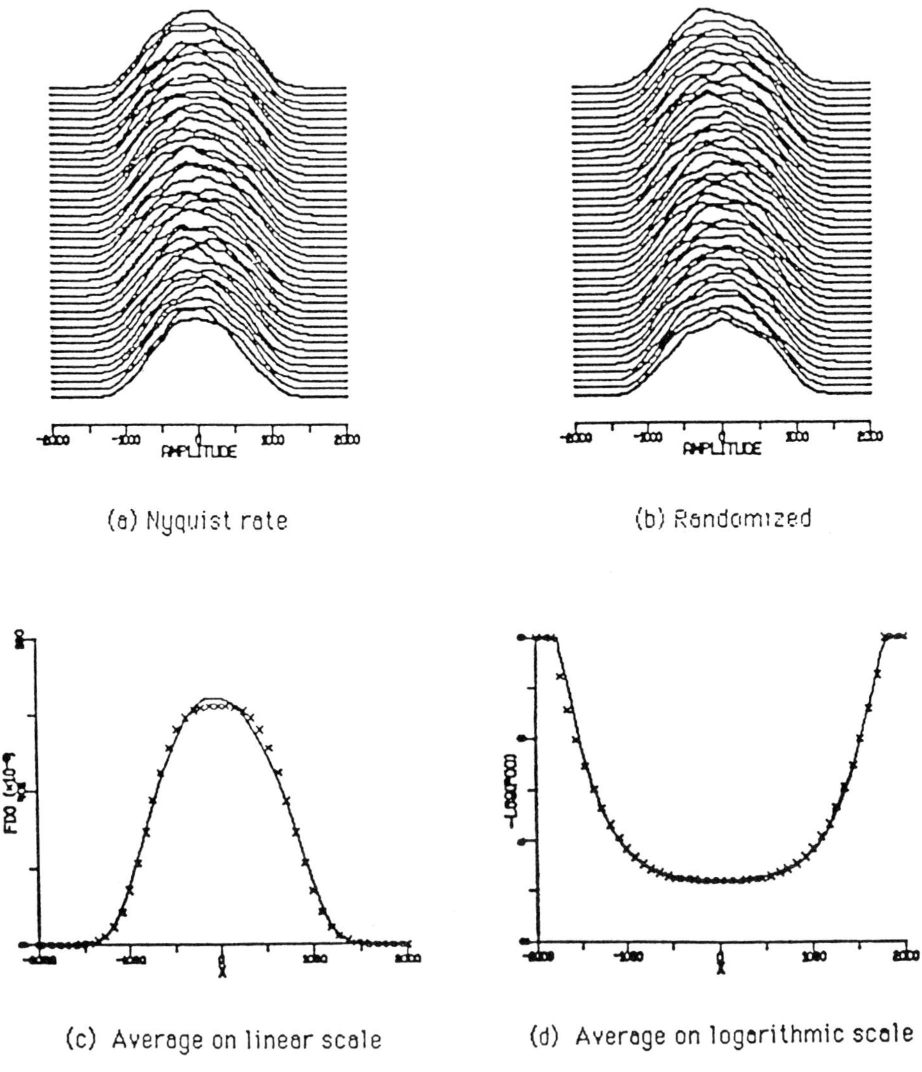

(a) Nyquist rate

(b) Randomized

(c) Average on linear scale

(d) Average on logarithmic scale

FIGURE 14

KERNEL DENSITY ESTIMATES FOR MERCHANT DATA

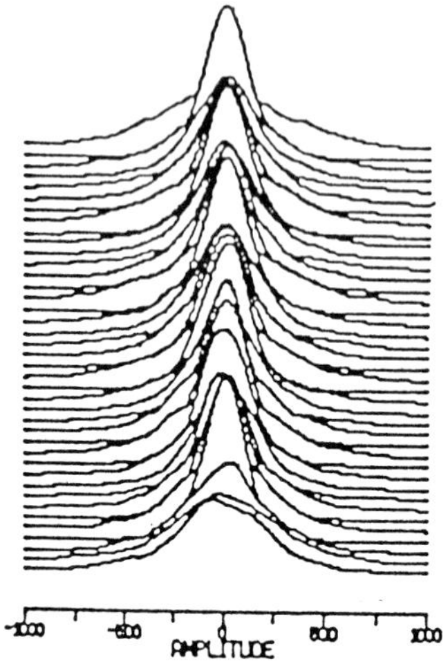

(a) Nyquist rate

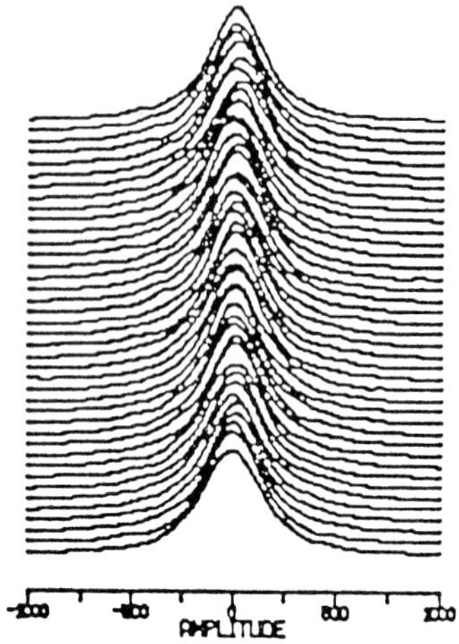

(b) Randomized

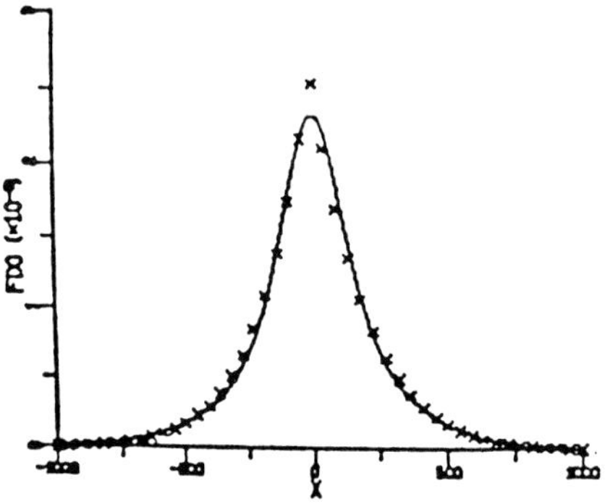

(c) Average on linear scale

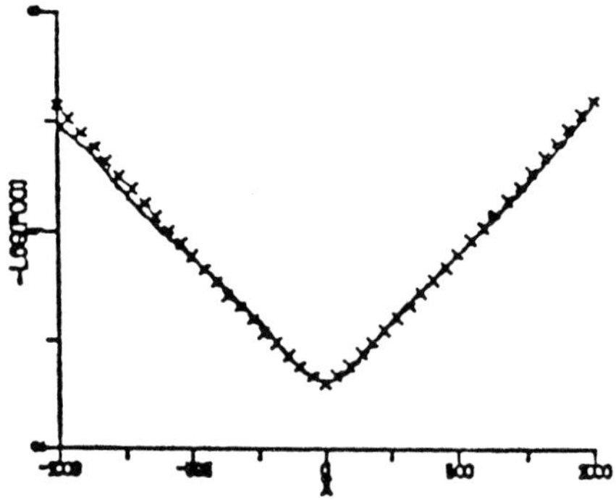

(d) Average on logarithmic scale

FIGURE 15

KERNEL DENSITY ESTIMATES FOR SEISMIC DATA

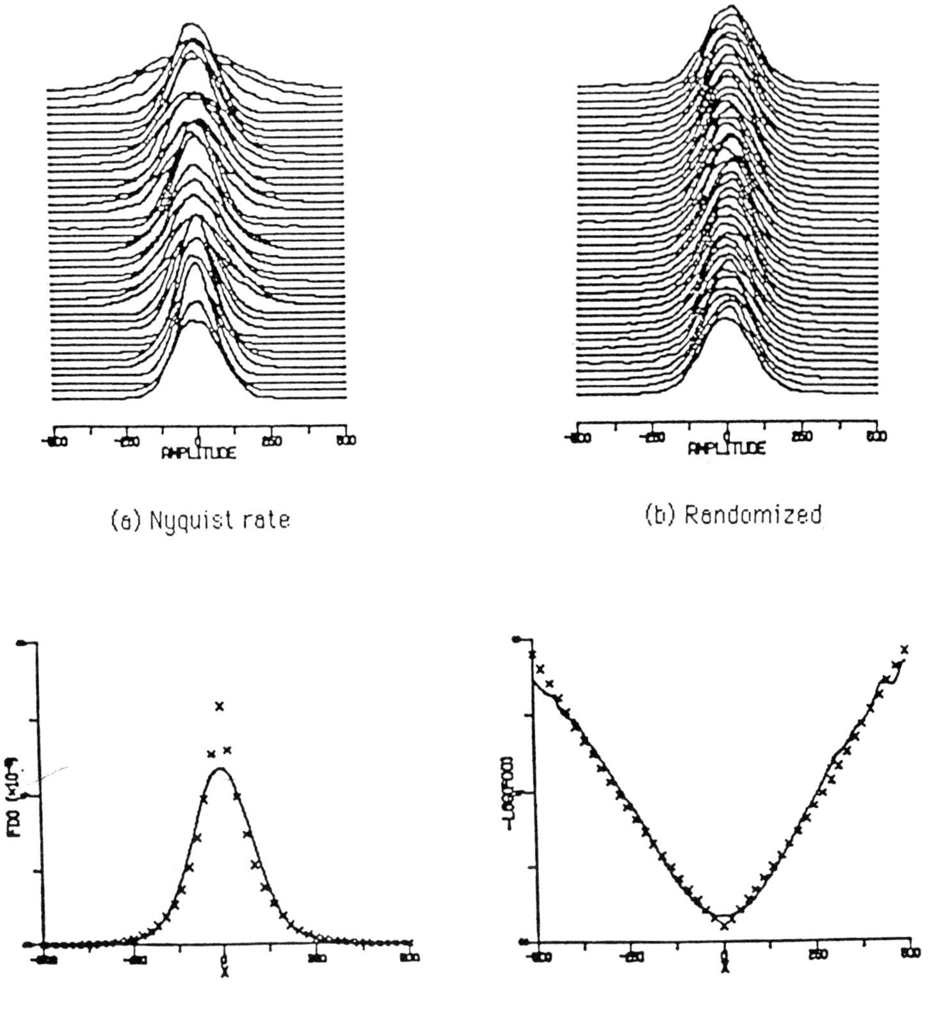

(a) Nyquist rate

(b) Randomized

(c) Average on linear scale

(d) Average on logarithmic scale

FIGURE 16

KERNEL DENSITY ESTIMATES FOR ANTARCTIC DATA

CONDITIONALLY LINEAR AND NON-GAUSSIAN PROCESSES[1]

R.R. Mohler[2] and W.J. Kolodziej
Department of Electrical and Computer Engineering
Oregon State University
Corvallis, OR 97331

ABSTRACT

A new theory and corresponding methodology is evolving for certain classes of non-linear non-Gaussian signal processing. This research considers particular statistical model structures such as conditionally linear or bilinear. Some non-Gaussian distributions, which arise in underwater acoustical signal processing, are included, as are others. The results are based on rigorous developments related to, but not limited to, bilinear and conditionally Gaussian processes. Parameter or state estimation (e.g., acoustic source location) are studied as well as preliminary results in information transmission and coding.

I. INTRODUCTION

Stochastic, partially observable systems, with linear-in-observable state dynamics are termed conditionally linear ("bilinear") systems here. It is well known that the solution of a state estimation problem for a conditionally linear system with Gaussian distribution of the initial state is given in terms of two sets of sufficient statistics, satisfying stochastic differential equations [1].

This result has been generalized [2] and applied successfully to acoustical signal processing for source location [3]-[5] which includes nonlinear, non-Gaussian structures.

Solved here is the state estimation problem which generalizes the above result for the case of an arbitrary a priori distribution. The method applied in this study is based on the derivation of an explicit formula for the conditional characteristic function of the state, given the past and present observations. This approach seems to impose less restrictive conditions on the system structure than the methods based on the derivation of the conditional distribution function.

It is shown here that the conditional characteristic function of the present and past states, given the present and past observations, is parametrically determined by a finite number of sufficient statistics. This result leads to the derivation of a processor, in the form of a finite set of stochastic differential equations which extends the result of [6] in a similar manner as a conditionally Gaussian filter generalizes a Kalman filter.

Then, in section 2, the results are applied to statistical coding and decoding for arbitrary (including non-Gaussian) processes.

[1] Supported by ONR Contract No. N00014-81-K0814
[2] R.R. Mohler is NAVELEX Chair Professor of Electrical and Computer Engineering, Naval Postgraduate School, Monterey, CA 93943, 1983-85.

1. Estimation and Statistical Description

Given the following system of stochastic differential equations

$$dx_t = (f_0(t,y) + f_1(t,y)x_t)dt + g_0(t,y)dw_t + q_0(t,y)dv_t, \tag{1.1}$$

$$dy_t = (h_0(t,y) + h_1(t,y)x_t)dt + dv_t, \quad 0 \le t \le T, \tag{1.2}$$

where f_0, f_1, g_0, q_0, h_0, h_1 are the nonanticipative functionals of y (i.e., $\mathbf{Y}_t$ measurable with $\mathbf{Y}_t = \sigma - alg\{y_s, 0 \le s \le t\}$), assuming that x_t, y_t satisfy (1.1) and (1.2), and that the conditional distribution of the initial states $F(a) = P(x_0 \le a \mid y_0)$ is given.

The organization of this section starts with Lemma 1, whereby it is shown that the conditional characteristic function of $(x_{t0}, x_{t1}, ..., x_{tn}) \mid \mathbf{Y}_t$, for an arbitrary decomposition $0 \le t_0 \le t_1 < \cdots < t_n \le t \le T$, of the interval [0,T] is of a particular form. Results from the theory of conditionally Gaussian processes are used here.

Next, Lemma 2, the explicit formula for the characteristic function of $x_t \mid Y_t$ is derived, and finally, in Lemma 3, all the results are organized to yield the recursive, finite-dimensional set of filter equations.

The assumptions used in the proof of Lemma 1 and 2 are listed below:

Let C_T denote the space of continuous functions $\eta = \{\eta_t, 0 \le t \le T\}$. It is assumed that for each $\eta \in C_T$

$$\int_0^T \Big(\sum_{k=0}^{1} (|f_k(t,\eta)| + |h_k(t,\eta)|) + |g_0(t,\eta)|^2 + |q_0(t,\eta)|^2\Big)\, dt < \infty \tag{1.3}$$

The above assumption assures the existence of the (Ito) integrals in (1.1) and (1.2) [7]. In order to use the results for conditionally Gaussian processes it is also assumed that [1]:

$$\text{for all } \eta \in C_T, \quad t \in [0,T], \quad |f_1(t,\eta)| + |h_1(t,\eta)| \le \text{const}, \tag{1.4}$$

and

$$\int_0^T E(|f_0(t,y)|^4 + |g_0(t,y)|^4 + |q_0(t,y)|^4)dt < \infty, \ \mathbf{E}(|x_0|^4) < \infty. \tag{1.5}$$

Lemma 1

Let

$$\phi_t = exp(i \sum_{k=0}^{n} z_k x_{tk}), \quad z = \begin{bmatrix} z_1 \\ \cdot \\ \cdot \\ \cdot \\ z_n \end{bmatrix} \in R^n, 0 \le t_0 < t_1 < \cdots < t_n \le t \le T.$$

Then the conditional characteristic function of $(x_{t0}, x_{t1}, \cdots, x_{tn}) \mid \mathbf{Y}_t$ is given by

$$\mathbf{e}_t(z) = \mathbf{E}(\phi_t \mid Y_t) = \int_{-\infty}^{\infty} exp(Q(t,a,z,y))\, dF(a) \tag{1.6}$$

where $Q(t,a,z,y)$ is quadratic in the variables a and z.

First notice that (1.1) solves as

$$x_t = \Phi_t(x_0 + \int_0^t \Phi_s^{-1}(f_0 - g_0 h_0)ds + \int_0^t \Phi_s^{-1} g_0 dy_s + \int_0^t \Phi_s^{-1} g_0 dw_s) \tag{1.7}$$

where

$$\Phi_t = exp(\int_0^t (f_1 - g_0 h_1)ds).$$

Rewrite (1.7) in the symbolic way as

$$x_t = \psi_t(x_0, w, y). \tag{1.8}$$

Now, the following version of the Bayes formula will be used [1, p. 8]: Let $\phi_t(x_0, w, y)$ be a nonanticipative functional of its arguments with $\mathbf{E}(\,|\phi_t\,|) < \infty$ for all $t \in [0,T]$. Then

$$E(\phi_t\,|Y_t) = \int_{-\infty}^{\infty} \int_{C_T} \phi_t(a,\eta,y)\rho_t(a,\eta,y)d\mu_w(\eta)dF(a) \tag{1.9}$$

where μ_w is a Wiener measure in the measurable space of continuous functions η on [0,T],

$$\rho_t(a,\eta,y) = exp(\int_0^t h_1(\psi_s(a,\eta,y) - \hat{x}_s(y))d\upsilon_s - \frac{1}{2}\int_0^t h_1^2(\psi_s(a,\eta,y) - \hat{x}_s(y))^2 ds) \tag{1.10}$$

with $d\upsilon_s = dy_s - (h_0 + h_1\,\hat{x}_s)ds$, and $\psi_s(a,\eta,y)$ defined by (1.8). The random process υ_t can be represented by

$$\upsilon_t = \int_0^t (dy_s - (h_0(s,y) + h_1(s,y)\hat{x}_s(y))ds) = v_t + \int_0^t h_1(s,y)(x_s - \hat{x}_s(y))ds.$$

Now using the Ito formula we have

$$e^{iz\upsilon_t} = e^{iz\upsilon_s} + iz\int_s^t h_1(\tau,y)e^{iz\upsilon_\tau}(x_\tau - \hat{x}_\tau(y))d\tau$$

$$+\, iz\int_s^t e^{iz\upsilon_\tau}dv_\tau - \frac{z^2}{2}\int_s^t e^{iz\upsilon_\tau}d\tau.$$

Multiplying both sides of the above equation by $e^{iz\upsilon_s}$ and taking the conditional expectation $\mathbf{E}(\cdot\,|\mathbf{Y}_s)$ gives

$$\mathbf{E}(e^{iz(\upsilon_t - \upsilon_s)}\,|\mathbf{Y}_s) = 1 - \frac{z^2}{2}\int_s^t \mathbf{E}(e^{iz(\upsilon_\tau - \upsilon_s)}\,|\mathbf{Y}_s)d\tau.$$

Solving the last equation yields

$$\mathbf{E}(e^{iz(\upsilon_t - \upsilon_s)}\,|\mathbf{Y}_s) = e^{-\frac{z^2}{2}(t-s)}, \tag{1.11}$$

which shows that $(\upsilon_t, \mathbf{Y}_t)$ is a Wiener process. Now rewrite $\rho_t(a,\eta,y)$ in a more convenient form. To this end introduce the following notation:

$$A_1(t,y) = h_1(\Phi_t(\int_0^t \Phi_s^{-1}(f_0 - q_0 h_0)ds + \int_0^t \Phi_s^{-1} q_0 dy_s) - \hat{x}_t),$$

$$A_2(t,y) = h_1\Phi_t,$$

$$A_3(t,y) = \Phi_t^{-1} g_0,$$

$$C_1(t,y) = \int_0^t A_1(s,y)dv_s - \frac{1}{2}\int_0^t A_1^2(s,y)ds ,$$

$$C_2(t,y) = \int_0^t A_2(s,y)dv_s - \int_0^t A_1(s,y)A_2(s,y)ds ,$$

$$C_3(t,y) = \int_0^t A_2^2(s,y)ds,$$

$$C_4(t,y,w) = \int_0^t A_2(s,y)\int_0^s A_3(\tau,y)dw_\tau dv_s - \int_0^t A_1(s,y)A_2(s,y)\int_0^s A_3(\tau,y)dw_\tau ds ,$$

$$C_5(t,y,w) = -\int_0^t A_2^2(s,y)\int_0^s A_3(s,y)dw_\tau ds .$$

Note also that $C_4(t,y,w)$ and $C_5(t,y,w)$ can be rewritten with the use of the Ito formula by:

$$C_4(t,y,w) = \int_0^t A_4(t,s,y)dw_s ,$$

$$C_5(t,y,w) = \int_0^t A_5(t,s,y)dw_s ,$$

where

$$A_4(t,s,y) = (\int_s^t A_2(\tau,y)d\nu_s - \int_s^t A_1(\tau,y)\, A_2(\tau,y)d\tau)\, A_3(s,y),$$

$$A_5(t,s,y) = -(\int_s^t A_2^2(\tau,y)d\tau)\, A_3(s,y) .$$

Now, using the above notation we have from (1.8) and (1.10)

$$\rho_t(a,w,y) = exp(C_1 + a(C_2 + C_5) + C_4 - \frac{a^2}{2}C_3 - \frac{1}{2}\int_0^t A_2^2(\int_0^s A_3 dw_\tau)^2 ds)$$

$$= exp(C_1 + aC_2 - \frac{a^2}{2}C_3 + \int_0^t (aA_5 + A_4)dw_s - \frac{1}{2}\int_0^t A_2^2(\int_0^s A_3 dw_\tau)^2 ds) . \tag{1.12}$$

The arguments in (1.12) were omitted for brevity. From (1.8) it follows that

$$x_t = \psi_t(x_0,w,y) = \Phi_t(x_0 + A_6(t,y) + \int_0^t A_3(s,y)dw_s)$$

where

$$A_6(t,y) = \int_0^t \Phi_s^{-1} (f_0 - q_0 h_0) ds + \int_0^t \Phi_s^{-1} q_0 dy_s .$$

Combining (1.12) and the above

$$\begin{aligned} exp(Q(t,a,z,y)) &= \int_{C_T} \phi_t(a,\eta,y)\, \rho_t(a,\eta,y) d\mu_w(\eta) \\ &= exp(C_1 + aC_2 - \frac{a^2}{2} C_3 + a(\sum_{k=1}^n \Phi_{tk} iz_k) \\ &+ \sum_{k=1}^n \Phi_{tk} A_6(t_k,y) iz_k) \int_{C_T} exp(\int_0^t (aA_5 + A_4) d\eta_s \\ &+ \sum_{k=1}^n iz_k \Phi_{tk} \int_0^{t_K} A_3 d\eta_s - \frac{1}{2}\int_0^t A_2^2 (\int_0^s A_3 d\eta_\tau)^2 ds) d\mu_w(\eta) . \end{aligned} \tag{1.13}$$

In order to evaluate the integral in (1.13) the following results will be used:

(i) Since the above integral represents a conditional expected value of its integrand, under the condition that y_s, $s \in [0,t]$ and $x_0 = a$ are given, the resulting distributions are of conditionally Gaussian type [1]. Note that this fact does not depend on the $F(a)$.

(ii) With all the variables in (1.13) being conditionally Gaussian we can use a convenient theorem:

Theorem [1, pp. 12-13]

Let w_t, $t \in [0,T]$ be a Wiener process and let $R(t)$, $G(t)$, and $H(t) \geq 0$ be such that

$$\int_0^T (|R(t)| + G(t)^2 + H(t)) dt < \infty .$$

Then for all $t \in [0,T]$

$$\begin{aligned} \mathbf{E}(exp(\int_0^t R(s)G(s)dw_s - \int_0^t H(s)(\int_0^s G(\tau)dw_\tau)^2 ds) \\ = exp(\frac{1}{2} D(t) + \frac{1}{2}\int_0^t G(s)^2\Gamma(s) ds) \end{aligned} \tag{1.14}$$

where

$$d\Gamma(s) = (2H(s) - \Gamma(s)^2 G(s)^2) ds ,\ \Gamma(t) = 0$$

and $D(t)$ is the covariance of $\int_0^t R(s) d\xi_s$, where

$$d\xi_s = G(s)\Gamma(s)\xi_s ds + G(s) dw_s ,\ \xi_0 = 0.$$

Comparing the last integral in (1.13) with the equation given by (1.14), we note that the corresponding $R(t)$ is a linear function of a and z. Now (1.9), (1.13), (1.14), and the definition of $D(t)$ conclude the proof of Lemma 1.

From Lemma 1 it follows in particular that for $z \in \mathbf{R}$, the characteristic function of $x_t \mid \mathbf{Y}_t$ is given by

$$\mathbf{e}_t(z) = C(t,y) \int_{-\infty}^{\infty} exp(a^2F_1(t,y) + aF_2(t,y) + izaF_3(t,y) + izF_4(t,y) + z^2F_5(t,y))dF(a) , \tag{1.15}$$

where F_1, F_2, F_3, F_4, F_5 do not depend on $F(a)$. Normalizing $\mathbf{e}_t(z)$ (i.e., requiring that $\mathbf{e}_t(0) = 1$) yields

$$\mathbf{e}_t(z) = exp(izF_4 + z^2F_5) \frac{\int_{-\infty}^{\infty} exp(a^2F_1 + a(F_2 + izF_3))dF(a)}{\int_{-\infty}^{\infty} exp(a^2F_1 + aF_2)dF(a)}. \tag{1.16}$$

Then from the general properties of the characteristic function, it follows that

$$\frac{1}{i} \frac{d\mathbf{e}_t(z)}{dz} \Big|_{z=0} = \hat{x}_t ,$$

$$(\frac{1}{i})^2 \frac{d^2\mathbf{e}_t(z)}{dz^2} \Big|_{z=0} = P_t + \hat{x}_t^2 ,$$

where $P_t = \mathbf{E}((x_t - \hat{x}_t)^2 \mid \mathbf{Y}_t)$, i.e., the conditional variance of $x_t \mid \mathbf{Y}_t$. From the above and (1.16)

$$\hat{x}_t = F_3 I_t(1) + F_4 , \tag{1.17}$$

$$P_t = -2F_5 + F_3^2 (I_t(2) - I_t^2(1)) , \tag{1.18}$$

where

$$I_t(n) = \frac{\int_{-\infty}^{\infty} a^n \, exp(a^2F_1 + aF_2)dF(a)}{\int_{-\infty}^{\infty} exp(a^2F_1 + aF_2)dF(a)} , \quad n = 1,2 . \tag{1.19}$$

The following Lemma defines F_i , $i = 1,2,3,4,5$ in (1.16).

Lemma 2

The characteristic function of $x_t \mid \mathbf{Y}_t$ is given by

$$\mathbf{e}_t(z) = exp(-\frac{1}{2} z^2\bar{P}_t(0)) \frac{\int_{-\infty}^{\infty} exp(a^2F_1 + aF_2 + iz\bar{x}_t(a,0))dF(a)}{\int_{-\infty}^{\infty} exp(a^2F_1 + aF_2)dF(a)} , \tag{1.20}$$

where $\bar{x}_t(a,0)$, $\bar{P}_t(0)$ are given as the solutions to the following set of differential equations with $\sigma = 0$:

$$d\bar{x}_t(a,\sigma) = (f_0 + f_1\bar{x}_t(a,\sigma))dt + (q_0 + \bar{P}_t(\sigma)h_1)(dy_t - (h_0 + h_1\bar{x}_t(a,\sigma))dt) \tag{1.21}$$

$$\bar{x}_0(a,\sigma) = a \ ,$$

$$d\bar{P}_t(\sigma) = (2f_1\bar{P}_t(\sigma) + g_0^2 + q_0^2 - (q_0 + \bar{P}_t(\sigma)h_1)^2)dt \ , \ \bar{P}_0(\sigma) = \sigma^2 \ , \tag{1.22}$$

and

$$F_1 = -\frac{1}{2}\int_0^t h_1^2\phi_s^2 ds \ , \tag{1.23}$$

$$F_2 = \int_0^t \phi_s h_1(dv_s + h_1\phi_s I_s(1)ds) \ , \tag{1.24}$$

$$\phi_t = exp(\int_0^t (f_1 - h_1(q_0 + \bar{P}_s(0)h_1))ds) \ . \tag{1.25}$$

Proof of Lemma 2

Since the F_i do not depend on $F(a)$ (see Lemma 1) take

$$dF(a) = \frac{1}{\sqrt{2\pi\sigma}} exp(-\frac{(a-m)^2}{2\sigma^2})da, \ a, \sigma > 0 \ . \tag{1.26}$$

In this case the resulting conditionally Gaussian distribution allows for explicit $\mathbf{e}_t(z)$ calculation [1]. Accordingly,

$$\mathbf{e}_t(z) = exp(iz\bar{x}_t(m,\sigma) - \frac{1}{2} z^2\bar{P}_t(\sigma)) \ , \tag{1.27}$$

where $\bar{x}_t(m,\sigma)$ and $\bar{P}_t(\sigma)$ satisfy (1.21) and (1.22) respectively. With $F(a)$ given by (1.26) it follows from (1.16) that

$$\mathbf{e}_t(z) = exp(iz(F_4 + \hat{\sigma}^2(F_2 + \frac{m}{\sigma^2})F_3) + z^2(F_5 - \frac{1}{2}\hat{\sigma}^2F_3^2)) \ , \tag{1.28}$$

where

$$\hat{\sigma}^{-2} = \sigma^{-2} - 2F_1 \ .$$

Comparing (1.27) and (1.28), we have

$$\bar{x}_t(m,\sigma) = F_4 + \hat{\sigma}^2(F_2 + \frac{m}{\sigma^2})F_3 \ , \tag{1.29}$$

and

$$\bar{P}_t(\sigma) = \hat{\sigma}^2F_3^2 - 2F_5 \ . \tag{1.30}$$

Letting now $\sigma \to 0$ in (1.29) and (1.30), it follows that

$$F_4 + mF_3 = \bar{x}_t(m,0) \ , \tag{1.31}$$

and

$$F_5 = -\frac{1}{2}\,\overline{P}_t(0)\,. \tag{1.32}$$

The above allows $\mathbf{e}_t(z)$ to be of the form of (1.20) with F_1 and F_2 yet to be defined. Using now (1.17) and (1.18) and explicitly calculating $I_t(n)$, $n = 1,2$,

$$\Delta_t(\sigma^{-2} - 2F_1) = F_3^2 \tag{1.33}$$

and

$$\Delta_t(\frac{m}{\sigma^2} + F_2) = F_3(\hat{x}_t - F_4)\,, \tag{1.34}$$

with

$$\Delta_t = P_t - \overline{P}_t(0)\,.$$

The formulae for F_1 and F_2 will be obtained by differentiating (1.33) and (1.34). However, before this is done recall from the theory of nonlinear filtering [7] that in general for x_t, y_t given as a solution to (1.1) and (1.2) $\hat{x}_t$, P_t satisfy

$$d\hat{x}_t = (f_0 + f_1\hat{x}_t)dt + (q_0 + P_t h_1)d\upsilon_t,\; \hat{x}^0 = \int_{-\infty}^{\infty} a dF(a)\,, \tag{1.35}$$

$$dP_t = (2f_1P_t + g_0^2 + q_0^2 - (q_0 + P_t h_1)^2)dt + h_1R_t\,d\upsilon_t,\; P_0 = \int_{-\infty}^{\infty} (a - \hat{x}_0)^2dF(a), \tag{1.36}$$

where

$$R_t = \mathbf{E}((x_t - \hat{x}_t)^3\,|\mathbf{Y}_t)\,.$$

Remark. Direct application of Eqs. (1.35) and (1.36) meets the difficulty of infinite coupling between the subsequent moments. From Eqs. (1.22), (1.35), and (1.36), and the fact that for conditionally Gaussian processes $R_t = 0$,

$$d\Delta_t = \Delta_t(2f_1 - h_1(2q_0 + h_1(P_t + \overline{P}_t(0))))dt\,,\; \Delta_0 = \sigma^2. \tag{1.37}$$

Now from (1.33) and (1.34) (upon differentiation) and using (1.35) and (1.37) it follows that

$$dF_1 = -\frac{1}{2}\,h_1^2F_3^2dt\,,\; F_1(0) = 0 \tag{1.38}$$

and

$$dF_2 = F_3h_1(d\upsilon_t + h_1F_3I_t(1)dt),\; F_2(0) = 0\,. \tag{1.39}$$

To define F_3, notice that Eq. (1.21) solves as

$$\overline{x}_t(a,\sigma) = \phi_t(a + \int_0^t \phi_s^{-1}(f_0 - h_0(q_0 + \overline{P}_s(\sigma)h_1))ds + \int_0^t \phi_s^{-1}(q_0 + \overline{P}_s(\sigma)h_1)dy_s)\,, \tag{1.40}$$

where

$$\phi_t = exp(\int_0^t (f_1 - h_1(q_0 + \overline{P}_s(\sigma)h_1))ds. \tag{1.41}$$

Comparing the above with (1.31) shows that $F_3 = \phi_t$ for $\sigma = 0$, which ends the proof of Lemma 2.

Lemma 3 below merely organizes all the results into the filter equations and the final form of the conditional characteristic function.

Lemma 3

Given the system (1.1) and (1.2) together with the a priori distribution $F(a) = P(x_0 \le a \mid y_0)$. The following are the filter equations (i.e., formulae of the recursive type, which calculate $\hat{x}_t = \mathbf{E}(x_t \mid \mathbf{Y}_t)$).

$$d\hat{x}_t = (f_0 + f_1\hat{x}_t)dt + (q_0 + P_t h_1)d\upsilon_t ,\ \hat{x}_0 = \int_{-\infty}^{\infty} a dF(a) , \tag{1.42}$$

$$d\upsilon_t = dy_t - (h_0 + h_1\hat{x}_t)dt , \tag{1.43}$$

$$P_t = \overline{P}_t + \phi_t^2(I_t(2) - I_t^2(1)) , \tag{1.44}$$

$$d\overline{P}_t = (2f_1\overline{P}_t + g_0^2 + q_0^2 - (q_0 + \overline{P}_t h_1)^2)dt ,\ \overline{P}_0 = 0 , \tag{1.45}$$

$$I_t(n) = \frac{\int_{-\infty}^{\infty} a^n exp(a^2F_1 + aF_2)dF(a)}{\int_{-\infty}^{\infty} exp(a^2F_1 + aF_2)dF(a)} ,\ n = 1,2 , \tag{1.46}$$

$$dF_1 = -\frac{1}{2} h_1^2\phi_t^2 dt ,\ F_1(0) = 0 , \tag{1.47}$$

$$dF_2 = \phi_t h_1(d\upsilon_t + \phi_t h_1 I_t(1)dt),\ F_2(0) = 0 , \tag{1.48}$$

$$d\phi_t = (f_1 - h_1(q_0 + \overline{P}_t h_1))\phi_t\, dt ,\ \phi_0 = 1 . \tag{1.49}$$

The characteristic function of $x_t \mid \mathbf{Y}_t$ is given by:

$$\mathbf{e}_t(z) = exp(iz(\hat{x}_t - \phi_t I_t(1)) - \frac{1}{2} z^2\overline{P}_t) \frac{\int_{-\infty}^{\infty} exp(a^2F_1 + a(F_2 + iz\phi_t))dF(a)}{\int_{-\infty}^{\infty} exp(a^2F_1 + aF_2)dF(a)} \tag{1.50}$$

1.2 *Special Cases*

Two special cases of Eqs. (1.1) and (1.2) result in significant simplification of the filter equations. The first case occurs when $g_0(t,y) = 0$, $0 \le t \le T$. From (1.7) it follows then that x_t is of the form

$$x_t = A_t(y)x_0 + B_t(y) .$$

Using the above equation in (1.2) we have the following estimation problem: Let x_0 be a random variable with distribution $F(a) = P(x_0 \le a \mid y_0)$. Assume that the observation process y_t, $0 \le t \le T$, admits a differential

$$dy_t = (h_0(t,y) + h_1(t,y)x_0)dt + dv_t ,$$

where the notation stays the same as in (1.2) and h_0, h_1 satisfy (1.3) and (1.4). From

Lemma 2 it follows now that the conditional characteristic function of x_0 given $\mathbf{Y}_t$ is of the form

$$\mathbf{e}_t(z) = \frac{\int_{-\infty}^{\infty} exp(a^2F_1 + aF_2 + iza)dF(a)}{\int_{-\infty}^{\infty} exp(a^2F_1 + aF_2)dF(a)}. \quad (1.51)$$

The above results from the fact that $dx_0 = 0$ replaces Eq. (1.1) implying $\overline{P}_t(0) = 0$ and $\overline{x}_t(a,0) = a$, as defined by Eqs. (1.21) and (1.22). Now

$$\frac{d\mathbf{e}_t(z)}{dz}\Big|_{z=0} = i\hat{x}_t ,$$

where $\hat{x}_t = \mathbf{E}(x_0 \,|\mathbf{Y}_t)$, combined with the general filter equations (1.42) ÷ (1.50) yields

$$\hat{x}_t = \frac{\int_{-\infty}^{\infty} a \; exp(-\frac{1}{2} a^2 \int_0^t h_1^2 ds + a \int_0^t h_1(dy_s - h_0 ds))dF(a)}{\int_{-\infty}^{\infty} exp(-\frac{1}{2} a^2 \int_0^t h_1^2 ds + a \int_0^t h_1(dy_s - h_0 ds))dF(a)} . \quad (1.52)$$

In particular if $dF(a) = \frac{1}{\sqrt{2\pi}\sigma_0} exp(-1 \frac{(a - m_0)^2}{2\sigma_0^2})da$, (1.52) results in

$$\hat{x}_t = \frac{m_0 + \sigma_0^2 \int_0^t h_1(dy_s - h_0 ds)}{1 + \sigma_0^2 \int_0^t h_1^2 ds} . \quad (1.54)$$

The above agrees with the result presented in [1, pp. 22-24].

The second special case for which the filter takes a simple form follows if $h_1(t,y) = 0$ (i.e., the state is not observable directly), $0 \le t \le T$. Now the filter equations (1.42) ÷ (1.50) reduce to

$$\begin{aligned} d\hat{x}_t &= (f_0 + f_1\hat{x}_t)dt + g_0(dy_t - h_0 dt) , \\ \hat{x}_0 &= \int_{-\infty}^{\infty} adF(a) , \\ dP_t &= (2f_1P_t + g_0^2)dt , \\ P_0 &= \int_{-\infty}^{\infty} (a - \hat{x}_0)^2 dF(a) . \end{aligned} \quad (1.55)$$

Separation of optimal estimation and decision is studied in [9].

2. Optimal Coding and Decoding for Transmission of an Arbitrary Signal

The results obtained for conditionally linear processes above and in [9] allow the study of optimization methods for transmission of an arbitrary process (including non-

Gaussian processes) through channels with additive white noise using noiseless feedback.

Assume first that the signal to be transmitted is a random variable θ with the distribution function

$$P(\theta \leq a) = F(a),$$

where $F(a)$ is known at both the transmitting and the receiving ends. The signal y_t, $t \in [0,T]$ at the transmitter output is assumed to satisfy the stochastic differential equation

$$dy_t = h(t,\theta,y)dt + dw_t, \; y_0 = 0, \tag{2.1}$$

where w_t, $0 \leq t \leq T$, is a Wiener process independent of θ. The nonanticipatory functional $h(t,\theta,y)$, determines the coding and is assumed to be such that the equation (2.1) has a unique solution.

Assume also that the functionals $h(\cdot)$ are subject to certain constraints. Here, for $0 \leq t \leq T$,

$$E(h_t^2) \leq \rho, \tag{2.2}$$

which has the interpretation of a finite average power. (ρ is a given constant.)

For each $t \in [0,T]$ the output signal $\hat{\theta}_t(y)$ can be constructed from the received signal $\{y_s, 0 \leq s \leq t\}$. The nonanticipative functional $\hat{\theta}_t(y)$ represents the decoding and must be chosen to reproduce the signal θ in the optimal manner. Here, optimal refers to

$$\inf_{\hat{\theta}} \mathbf{E}(\theta - \hat{\theta}_t)^2, \quad 0 \leq t \leq T, \tag{2.3}$$

since, with given coding $\mathbf{E}(\theta - \hat{\theta}_t(y))^2 \geq \mathbf{E}(\theta - \hat{x}_t(y))^2$, where $\hat{x}_t(y) = \mathbf{E}(\theta \,|\, \mathbf{Y}_t = \sigma - alg\{y_s, 0 \leq s \leq t\})$, the optimal decoding of the signal y is the posteriori mean.

In order to implement the above optimal decoding, consider the subclass of coding functionals $h(t,\theta,y)$ which depend linearly on θ:

$$h(t,\theta,y) = h_0(t,y) + h_1(t,y)\theta, \tag{2.4}$$

where h_0, h_1 are nonanticipative functionals.

$$\Delta_t = \inf_{h_0,\, h_1} \mathbf{E}(\theta - \hat{x}_t(y))^2 \tag{2.5}$$

The problem is to find the optimal coding functionals $\hat{h}_0$, $\hat{h}_1$ for which the infimum in (2.5) is attained. Hence, let some coding (h_0, h_1) be chosen, and let y_t satisfy the equations (2.1) and (2.4); then from (1.50) the conditional characteristic function of $\theta \,|\, y_t$ is given by

$$e_t(z) = \frac{\int_{-\infty}^{\infty} exp(-\frac{1}{2} a^2 \int_0^t h_1^2 ds + a \int_0^t h_1(dy_s - h_0 ds) + iza)dF(a)}{\int_{-\infty}^{\infty} exp(-\frac{1}{2} a^2 \int_0^t h_1^2 ds + a \int_0^t h_1(dys - h_0 ds))dF(a)}. \tag{2.6}$$

Furthermore,

$$P_t = E((\theta - \hat{x}_t)^2 \,|\, \mathbf{Y}_t) = -\left.\frac{d^2 e_t(z)}{dz^2}\right|_{z=0} - \left(\left.\frac{de_t(z)}{dz}\right|_{z=0}\right)^2 \tag{2.7}$$

and the following optimal-estimation algorithm results:

$$d\hat{x}_t = P_t h_1 d\upsilon_t,\; \hat{x}_0 = \int_{-\infty}^{\infty} a dF(a)\,,$$

$$P_t = I_t(2) - I_t^2(1)\,,$$

$$I_t(n) = \frac{\int_{-\infty}^{\infty} a^n\, exp\,(a^2 F_1 + aF_2) dF(a)}{\int_{-\infty}^{\infty} exp\,(a^2 F_1 + aF_2) dF(a)}\,, \tag{2.8}$$

$$dF_1 = -\frac{1}{2} h_1^2 dt,\; F_1(0) = 0\,,$$

$$dF_2 = h_1(d\upsilon_t + h_1 I_t(1)dt),\; F_2(0) = 0\,,$$

$$d\upsilon_t = dy_t - (h_0 + h_1\hat{x}_t)dt$$

Note from (2.5), that the optimal coding problem can be stated as follows:

$$\Delta_t = \inf_{h_0, h_1} E(P_t)\,. \tag{2.9}$$

From the general theory of nonlinear filtering,

$$dP_t = -h_1^2 P_t^2 dt + h_1 Q_t\, d\upsilon_t\,,$$

where $Q_t = \mathbf{E}((\theta - \hat{x}_t)^3 \,|\, Y_t)$. Consequently, the above yields

$$P_t = P_0 \exp\left(-\int_0^t h_1^2 P_s\, ds\right) + \int_0^t h_1 Q_s \exp\left(-\int_s^t h_1^2 P_s\, ds\right) d\upsilon_s\,.$$

Taking the expectation results in

$$\mathbf{E}(P_t) = E\left(P_0 \exp(-\int_0^t h_1^2 P_s\, ds)\right) \geq P_0 \exp\left(-\int_0^t E(h_1^2 P_s) ds\right) \tag{2.10}$$

In the above the fact that υ_t possesses all the properties of a Wiener process, and Jensen's inequality are used.

Taking into account the constraint (2.2) it is seen that

$$\mathbf{E}(h_t^2) = \mathbf{E}(h_0 + h_1\theta)^2 = \mathbf{E}(h_0 + h_1(\theta - \hat{x}_t) + h_1\hat{x}_t)^2 = \tag{2.11}$$
$$= \mathbf{E}(h_0 + h_1\hat{x}_t)^2 + \mathbf{E}(h_1^2 P_t) \leq \rho\,.$$

Now from (2.10), it follows that

$$\mathbf{E}(P_t) \geq P_0 \exp(-\rho \cdot t)\,. \tag{2.12}$$

For the optimal coding the inequality in (2.12) has to be replaced with equality. This will occur if

$$\hat{h}_1 = \sqrt{\frac{\rho}{P_t}} . \tag{2.13}$$

The above implies from (2.11) that

$$\mathbf{E}(\hat{h}_0 + \hat{h}_1 \hat{x}_t)^2 = 0$$

which is accomplished if and only if, $\hat{h}_0 = -\hat{h}_1 \hat{x}_t$. Hence the optimal coding takes place if the transmitted signal $\hat{y}_t$ satisfies the equation

$$d\hat{y}_t = \sqrt{\frac{\rho}{P_t}} (\theta - \hat{x}_t)dt + dw_t ,\ \hat{y}_0 = 0 , \tag{2.14}$$

showing that only the transmission of the difference between the value of θ and its optimal estimate $\hat{x}_t$ multiplied by $\sqrt{\rho / P_t}$ occurs.

Note that for the optimal coding $\hat{y}_t$ coincides with the innovation process,

$$v_t = \int_0^t (d\hat{y}_s - (\hat{h}_0 + \hat{h}_1 \hat{x}_s)ds) .$$

Consequently, in the optimal case the transmission is such that only the innovation process has to be transmitted.

Remark: In order for $\hat{h}_1$ (see (2.13)) to be well defined it is enough to show that $P_t > 0$ a.e. This can be shown, assuming that $P_0 > 0$, from the filter equations (2.8).

It can be shown also that the coding ($\hat{h}_0$, $\hat{h}_1$) defined by (2.14) has the largest information $I_t(\theta,y)$ about θ in the received signal y_s, $s \in [0,t]$ for each $t \in [0,T]$. Here the 'standard' definition of $I_t(\theta,y)$ is assumed i.e.,

$$I_t(\theta,y) = \mathbf{E}\left\{ \ln \frac{d\,\mu_{\theta,y}}{d\,[\mu_\theta \times \mu_y]} (\theta,y) \right\} . \tag{2.15}$$

The measures $\mu_{\theta,y}$, μ_θ and μ_y are associated with the processes (θ,y), θ and y respectively, and $\mu_\theta \times \mu_y$ is a Cartesian product of the measures μ_θ and μ_y. It is also assumed that $I_t(\theta,y) = \infty$ if $\mu_{\theta,y}$ is not absolutely continuous with respect to $\mu_\theta \times \mu_y$. It can be shown that (2.15) reduces, for y satisfying (2.1), to

$$I_t(\theta,y) = \frac{1}{2} \int_0^t \mathbf{E}(h^2(s,\theta,y) - \hat{h}_s^2(y))ds , \tag{2.16}$$

where $\hat{h}_t(y) = E(h(t,\theta,y)\,|\mathbf{Y}_t)$. From the above and the constraint (2.2)

$$I_t(\theta,y) \le \frac{1}{2} \int_0^t \mathbf{E}(h^2(s,\theta,y))dt \le \frac{1}{2} \rho \cdot t . \tag{2.17}$$

On the other hand, with optimal linear coding

$$h(t,\theta,y) = \sqrt{\frac{\rho}{P_t}} (\theta - \hat{x}_t) ,$$

and

$$\hat{h}_t(y) = 0,\ Eh^2 + E(\frac{\rho}{P_t} \cdot P_t) = \rho \ .$$

Hence from (2.16) it follows that $I_t(\theta,y) = \frac{1}{2}\rho \cdot t$. Comparing this with (2.17) proves that the linear coding which minimizes the transmission error also maximizes the amount of received information.

3. Conclusions and Projections

Optimal state estimation, coding, and decoding is derived for a nonlinear class of processes, termed conditionally linear or "bilinear." Non-Gaussian statistics are considered, and the results represent new contributions beyond the linear Gaussian, or previously considered nonlinear, statistical theory. The results show promise for acoustical signal-processing applications in an ocean setting. In this sense, the previously developed methods [3]-[5] can be generalized. Also, further related research is in progress on signal processing aspects of information content, stochastic observability, Lie-algebraic structures and adaptive features.

REFERENCES

1. R.S. Liptser and A.N. Shiryayev, *Statistics of Random Processes II - Applications,* Springer-Verlag, New York, 1978.

2. R.R. Mohler and W.J. Kolodziej, "Optimal Control of a Class of Nonlinear Stochastic Systems," *IEEE Trans. Auto. Cont.,* Vol. AC-26, pp. 1048-1053, 1981.

3. T.U. Halawani, R.R. Mohler and W.J. Kolodziej, "A Two-Step Bilinear Filtering Approximation," *IEEE Trans. Acous., Sp. & Sig. Proc.,* Vol. ASSP-32, pp. 244-352, 1984.

4. R.R. Mohler, W.J. Kolodziej, H.D. Brunk and R.S. Engelbrecht, "On Nonlinear Filtering and Tracking," in *Signal Processing in the Ocean Environment,* (E.J. Wegman, Ed.), Marcel-Dekker, New York, 1984.

5. W.J. Kolodziej and R.R. Mohler, "Analysis of a New Nonlinear Filter and Tracking Methodology," *IEEE Trans. Infor. Theory,* Vol. IT-29, 1984, to appear.

6. V. Benes and I. Karatzas, "Estimation and Control for Linear, Partially Observable Systems with Non-Gaussian Initial Distribution," *Stochastic Processes and Their Applications,* Vol. 14, pp. 233-248, 1983.

7. R.S. Liptser and A.N. Shiryayev, *Statistics of Random Processes I - General Theory,* Springer-Verlag, New York, 1977.

8. M.H.A. Davis, "The Separation Principle in Stochastic Control via Girsanov Solutions," *SIAM J. Control,* Vol. 14, pp. 176-188, 1976.

9. W.J. Kolodziej and R.R. Mohler, "State Estimation and Control of Conditionally Linear Systems," to appear, *SIAM J. Control & Optimization,* Vol. 25, 1985.

A GRAPHICAL TOOL FOR DISTRIBUTION AND CORRELATION ANALYSIS OF MULTIPLE TIME SERIES

by

Edward J. Wegman
Center for Computational Statistics and Probability
George Mason University
Fairfax, VA

and

Chris Shull
Department of Decision Sciences
The Wharton School
University of Pennsylvania
Philadelphia, PA

ABSTRACT

This paper proposes the use of the parallel coordinate representation for the representation of multivariable statistical data, particularly for data arising in the time series context. We discuss the statistical interpretation of a variety of structures in the parallel coordinate diagrams including features which indicate correlation and clustering. One application is to graphically assess the finite dimensional distribution structure of a time series. A second application is to assessing structure of multichannel time series. An example of this latter application is given. It is shown that the parallel coordinate representation can be exploited as a graphical tool for using beamforming for short segments of ocean acoustic data.

I. INTRODUCTION

Traditional signal processing based on the additive Gaussian noise assumption has made light use of contemporary data analysis techniques. This is essentially as a consequence of the heavy structural assumptions made by such models and their relative effectiveness in settings with high signal to noise ratios. As the signal-to-noise ratio drops, however, detection and estimation techniques become more sensitive to deviations from distributional assumptions. Consequently, there is strong reason to explore the finite-dimensional distribution structure of signal and noise processes. The papers by Machell and Penrod (1984) and Wilson and Powell (1984) explore the one-dimensional structure of ocean acoustic signals and noise using data analysis techniques and effectively demonstrate the non-Gaussian character of the univariate distributions. The attempt to develop non-Gaussian signal processing techniques began in a coordinated way in 1980. The volumes by Wegman and Smith (1984) and Baker (1986) document much of this development. Nonetheless, the ability to explore the finite dimensional distribution structure as well as the multiple time series output of a sensor array is hampered by the lack of any serious data analytic tool for displaying graphically the structure of multivariable data.

In this paper, we introduce the notion of a parallel coordinate representation of high dimensional data vectors as a tool for data analysis of time series. The parallel coordinate representation was introduced originally by Inselberg (1985) as a tool in computational geometry (e.g. determine whether a point is interior to a convex region in hyperspace) and by Wegman (1986) as a data analytic tool. In this paper, we focus on applications to time series generated from signal/noise processes.

The parallel coordinate representation is, in effect, a general n dimensional scatter diagram. The classic scatter diagram is a fundamental tool in the construction of a model for data. It allows the eye to detect structure in data such as correlation, clustering, outliers and the like. The usual Cartesian coordinate diagram does not generalize beyond three dimensions easily because of the orthogonality requirements on the axes. It is simply not possible to have four or more orthogonal coordinates in our apparently three-dimensional world. The parallel coordinates idea then is simply to give up trying to preserve this orthogonality and, thus, to draw the coordinate axes as co-planner parallel coordinates. This leads to some rather elegant mathematical underpinnings based on projective geometry as we shall see.

Alternatives to scatter plots have been proposed by several authors. Chernoff (1973) proposed the use of faces as a technique for representing multivariate data. This technique is based on the observation that humans recognize subtle differences in faces and, thus, if one can make features on a face correspond to values of the data vector in the various dimensions, one can distinguish fairly subtle differences between observations. Similarly, the star diagram technique reported on by Fienberg (1979) represents a n-dimensional vector as a n-pointed star. The distance of each point from the center is proportional to the magnitude of the corresponding coordinate. The irregularly shaped stars can thus be compared for similarities or dissimilarities. The principal objection to each of these tools is that they are primarily oriented to making comparisons between individual items and do not convey very well the overall structure of the data set.

More recently, interest has focused on scatter diagram matrices. See for example Cleveland and McGill (1984) and Carr et al. (1986). With this method, an array of scatter diagrams is arranged in matrix form: that is to say, the i,j entry into the matrix is the scatter diagram of X_i versus X_j where the sample observations are random vectors $X = (X_1, \ldots, X_n)$. Of course, the j,i entry is the scatter diagram of X_j versus X_i so that both symmetries are shown. Normally, the diagonal matrices are left out. This is quite an effective tool, but is still essentially limited to a collection of two-dimensional pairwise comparisons. It is still difficult to gain a real sense of hyperdimensional structure.

II. PARALLEL COORDINATES

We propose as a high-dimensional data analysis tool the following representation. A vector $(X_1, \ldots, X_n)$ is plotted by plotting X_1 on axis 1, X_2 on axis 2 and so on through X_n on axis n. The points plotted in this manner are joined by a broken line. Figure 2.1 illustrates two points (one solid, one dashed) plotted in the parallel coordinate representation. In this illustration the two points agree in the fourth coordinate. A principle advantage of this plotting device is clear. Each vector, $(X_1, \ldots, X_n)$, is represented in a planar diagram so that each vector component has essentially the same representation. Moreover, since all vectors can be represented in the same diagram, the point pairwise comparison limitation of Chernoff faces and star diagrams is eliminated.

The parallel coordinate representation enjoys some elegant duality properties with the usual Cartesian orthogonal coordinate representation. Consider line l in the Cartesian coordinate plane given by

$$l: \ y = mx + b$$

and consider any two arbitrary points lying on that line, say $(a, ma + b)$ and $(c, mc + b)$. For simplicity of computation we consider the xy Cartesian axes mapped into the xy parallel axes as described in Figure 2.2. We superimpose a Cartesian coordinate axes t, u on the xy parallel axes so that the x parallel axis has the equation $u = 0$ in the tu plane and the y parallel axis has the equation $u = 1$. The point $(a, ma + b)$ in the xy coordinate axes maps into the line joining $(a,0)$ and $(ma + b,1)$ in the tu coordinate axes. Similarly $(c, mc + b)$ maps into the line joining $(c,0)$ and $(mc + b,1)$. It is a straightforward computation to show that these two lines intersect at a point given by $\bar{l}$: $(b/1-m, 1/1-m)$. Notice this point in the parallel coordinate plot depends only on m and b the parameters of the original line in the Cartesian plot. Thus $\bar{l}$ is the dual of l and we have the interesting duality result that points in Cartesian coordinates map into lines in parallel coordinates while lines in Cartesian coordinates map into lines in parallel coordinates.

For $0 < 1/1-m < 1$, m is negative and the intersection occurs between the parallel coordinate axes. For $m = -1$, the intersection is exactly midway. A ready statistical interpretation can be given. For highly negatively correlated data pairs, the dual segments in parallel coordinates will tend to cross near a single point between the two parallel coordinate axes. The scale of one of the variables may be transformed in such a way that the intersection occurs midway between the two parallel coordinate axes in which case the slope of the linear relationship is negative one.

In the case that $1/1-m < 0$ or $1/1-m > 1$, m is positive and the intersection occurs external to the region between the two parallel axes. In the special case $m = 1$, this formulation breaks down. However, it is clear that the point pairs would have been $(a, a + b)$ and $(c, c + b)$. The dual lines to these points would be described by the lines in the parallel coordinate space with slope $1/b$ and intercepts $-a/b$ and $-c/b$ respectively. Thus the duals of these points in the parallel coordinate space are parallel lines with slope $1/b$. We thus append the ideal points to the parallel coordinate plane to obtain a projective plane. These parallel lines intersect at the ideal point in direction $1/b$. In the statistical setting we have the following interpretation. For highly positively correlated data, we will tend to have lines not intersecting between the parallel coordinate axes. By suitable linear rescaling of one of the variables, the lines may be made approximately parallel in direction with slope $1/b$. In this case the slope of the linear relationship between the rescaled variable is one. See Figures 2.3a and 2.3b for an illustration of large positive and large negative correlations. These are diagnostic templates. Of course, non-linear relationships will not respond to a simple linear rescaling. See Figure 2.3c. However, by suitable non-linear transformations, it should be possible to transform to the positive correlation template (parallel lines) or the negative correlation template (pencil of lines or cross-over effect).

The point-line, line-point duality extends to conics. A more extensive discussion of the conic-conic duality is given in Dimsdale (1984) and Wegman (1986). The key notion from a statistical perspective is the transformation of an elliptical region. An ellipse in the Cartesian plane is mapped by the projective transformation to a locus of lines, a so-called line conic. Indeed this line conic is a line hyperbola. The line hyperbola has as its envelope a point hyperbola as illustrated in Figure 2.4.

As a matter of interest, there are several other surprising dualities. A point of inflection in Cartesian coordinates maps into a cusp in parallel coordinates. This is also a rather useful duality since points of inflection are rather difficult to detect visually while cusps are easy. A translation in Cartesian coordinates maps into a rotation in parallel coordinates while a rotation in Cartesian coordinates maps into a translation. This latter fact is also a rather useful fact since a rotation is a nonlinear transform (involving trigonometric functions) while a translation involves only additions and multiplications. Thus it is computationally more efficient to carry out a translation in the parallel coordinate space rather than the corresponding rotation in Cartesian coordinates. The transformation between the two involves a matrix multiplication so that it is not really practical to transform for efficiency alone unless a large number of rotations are needed.

III. DATA ANALYTIC INTERPRETATIONS

Since ellipses map into hyperbolas, we can see an easy template for diagnosing uncorrelated data set. We would expect the two dimensional scatter diagram to fall substantially within the boundary of the circumscribing circle. As illustrated in Figure 3.1a, the parallel coordinate representation would approximate a figure with hyperbolic envelope. As the correlation approached a negative one, the hyperbolic envelope would deepen so that in the limit we would have the pencil of lines, the so-called cross-over effect. As the correlation approached a positive one, the hyperbolic envelope would widen (with fewer and fewer crossovers) so that in the limit we would have parallel lines. Griffen (1958) invented as a graphical device for computing Kendall tau which is, in effect, a parallel coordinate diagram based on ranks. The computational formula is

$$\tau = 1 - \frac{4X}{n(n-1)}$$

where X is the number of intersections resulting by connecting the two rankings of each observation pair by lines, one ranking having been put in natural order. While Griffen's original formulation was framed in terms of ranks for both x and y axes, it is clear that the number of crossings is invariant to any monotone increasing transformation in either x or y, the ranks being one such transformation. Because of this scale invariance, one would correctly expect rank-based statistics to have an intimate relationship with parallel coordinates.

So far we have focused on primarily pairwise parallel coordinate relationships. The idea however is that we can, so to speak, stack these diagrams and represent all n-dimensions simultaneously. Figure 3.2 thus illustrates a 5-dimensional hypersphere plotted in parallel coordinates. A 5-dimensional ellipsoid would have a similar general shape but with hyperbolas of different depths. Of course, observations from a 5-dimensional Gaussian density with a covariance matrix of the form σ^2 I would have an approximately spherical 5-dimensional scatter plot, thus Figure 3.2 would be a diagnostic template for this situation. It should be fairly clear what the parallel coordinate representation of a multivariate normal data set should be.

Figure 3.3 is illustrative of some data structures one might see in a 5-dimensional data set. First it should be noted that the plots along any given axis represent dot diagrams, hence convey graphically the one dimensional marginal distributions. In this illustration, the first axis is meant to have an approximately normal distribution shape while axis two the shape of the negative of a chi-square. As discussed above, the pairwise comparisons can be made. Figure 3.3 illustrates a number of instances of linear (both negative and positive), nonlinear and clustering situations. Indeed, it is clear that there are two 3-dimensional clusters along coordinates 3, 4 and 5.

Consider finally the appearance of a mode in parallel coordinates. The mode is, intuitively speaking, the location of the most intense concentration of probability. Hence, in a sampling situation it will be the location of the most intense concentration of observations. Since observations are represented by broken line segments, the mode in parallel coordinates will be represented by the most intense bundle of broken line paths in the parallel coordinate diagram. Roughly speaking, we should look for the most intense flow through the diagram. In Figure 3.3 such a flow begins near the center of coordinate axis 1 and finishes on the left-hand side of axis 5.

Figure 3.3 thus illustrates data analysis features of the parallel coordinate representation including the ability to diagnose one-dimensional characteristics (marginal densities), two-dimensional features (correlations), three-dimensional features (clustering) and a five-dimensional feature (the mode).

In Figure 3.3, the parallel coordinate axes were ordered from 1 through 5. This allowed an easy comparison of 1 with 2, 2 with 3 and so on. The comparison of 1 with 3, 2 with 5 and so on is not easily done since these axes are not adjacent. A natural question arises about the number of permutations required so that in some permutation, every axis is adjacent to every other axis. Although there are $n!$ permutations, many of these duplicate adjacencies. Actually, far fewer are required.

Wegman (1986) gives a construction for determining the axes ordering. The first ordering is given by

$$n_{k+1} = (n_k + (-1)^{k+1}k) \ mod \ n \quad , \ k = 1,2,...,n-1$$

with $n_1 = 1$ where n is the dimension of the data vector. Subsequent orderings are determined by

$$n_{k+1}^{(j+1)} = (n_k^{(j)} + 1) \ mod \ n \quad , \ j = 1,2,...,[\frac{n-1}{2}]$$

where $[\cdot]$ is the greatest integer function and $n_k^{(1)} = n_k$. Thus $[\frac{n+1}{2}]$ is the minimal number of permutations needed to guarantee all adjacencies.

It is worth making the observation that for conventional scatter diagrams, $\binom{n}{2}$ presentations are needed to represent every pairwise comparison whereas only $[\frac{n+1}{2}]$ presentations are needed in parallel coordinates. Thus even if the parallel coordinate representation were not able to capture higher dimensional structure, it would still be a more economical representation.

IV. A DATA EXAMPLE AND SOME CONCLUSIONS

Figures 4.1 and 4.2 represent multichannel ocean acoustic time series data plotted in parallel coordinates. The interpretation of such diagrams takes some experience, but experience accumulated relatively easily. In both diagrams the measurements derive from a horizontal acoustic line array. The array elements were equally spaced except for that of channel 6 which was remote from the remaining elements. This anomalous property of channel 6 shows up readily in Figure 4.1. The data in channel 6 is skewed to the right relative to the other channels with several strong outliers on the low side. The remaining channels show no obvious correlation structures. Indeed, the approximate hyperbolic boundary is evident in the remaining pairs. This data is an example of an ambient deep water acoustic noise time series and essentially shows the uncorrelated approximately Gaussian noise we might expect.

Figure 4.2 in contrast is noise with a coherent broad-band signal present. This data plot is based on only 100 observations to show the ability to represent structure in highly noisy data of short integration time. The cross-over effect indicating negative correlation is among the easiest diagnostic templates to recognize. It is apparent in Figure 4.2 as the crossovers or, if preferred, as the diamond-shaped patterns. Lags were introduced into the data and adjusted until the patterns became evident. Unfortunately, our graphics tools were not interactive, consequently, this was a tedious process and thus we did not conduct extensive experimentation to guarantee these were optimal lags. The principle is clear, however. Even in a very low signal to noise ratio environment with broadband signals, it is possible to detect correlation structure graphically. Our graphical technique does not suffer from the conventional correlation technique's lack of robustness to distributional assumptions. This application is, of course, to time delay (lag) estimation for coherent signals, thus allowing beam steering. This is a visually oriented tool for doing beamforming.

A high-interaction implementation would allow very rapid assessment of correlation structure. For example, the lag of a particular axis could be tied to the position of a mouse. Movement of the mouse could dynamically change the lag which would be visually matched to some template. A click of the button on the mouse would change the axis being lagged. Thus in short order optimal lags could be determined.

The parallel coordinate diagram could be used as well for studying finite-dimensional process structure. Baker and Gualtierotti (1986), for example, consider signal detection for non-Gaussian processes based on likelihood ratios. Finite dimensional distributions are the basis for computation of the likelihood ratio. Thus, exploratory analysis of finite dimensional distributions is key to implementation of such likelihood based techniques.

Wegman (1986) discusses another example of data analysis via parallel coordinates as well as developing more fully the basic theory.

V. STRUCTURAL INFERENCE

The ultimate goal of a graphical analysis of the sort described here is to discover structure in the data. The obvious ability of the parallel coordinate diagram to represent correlations and distributional features makes it an excellent tool for dealing with these properties in a time series. While knowledge of correlation and distributional properties is of interest in its own right, a larger goal is to make inference about the structural relationships between random variables, the problem we call the structural inference problem.

Just as in an ordinary regression setting, we would like to determine the regression equation, $E(Y|X=x)=x\beta$, we should like to determine, in the nonlinear, high dimensional setting, the function f for which $E\, f(X_1, \ldots, X_n)=0$. This f represents the structural relationship between the random variables of interest (which, in fact, could be elements of a time series). We could, for example, be interested in computing $E_{n-1}\, f(X_1, \ldots, X_{n-1}, X_n)=0$ where E_n is the conditional expectation given the data up to time n, so that $E_{n-1} f(X_1, \ldots, X_n) = 0$ is the prediction equation for a non-linear time series model.

In general, the structure $\{(X_1, \ldots, X_n) \in E^n\colon E f(X_1, \ldots, X_n)\}$ describes a so-called algebraic variety in n-dimensional space. With fairly modest regularity conditions, an algebraic variety becomes a manifold topologically homeomorphic to Euclidean k-space, E^k, for some $k \leq n$. Of course, what we actually observe is not $\{(X_1, \ldots, X_n)\colon f(X_1, \ldots, X_n) = 0\}$ but $\{(X_1, \ldots, X_n)\colon f(X_1, \ldots, X_n) = \epsilon\}$, i.e., we observe scatter off of the manifold. Presumably if there is a real structural relationship, $M = \{(X_1, \ldots, X_n)\colon E f(X_1, \ldots, X_n) = 0\}$ will be a manifold of dimension k considerably smaller than n. In the simple regression setting $M = \{(X,Y)\colon E_X(Y) = X\beta\}$, M is one dimensional (a straight line) while the data is actually a two-dimensional point cloud.

In general, then, structural inference involves estimating M. A priori, we do not know the dimension of M. In essence, we propose here to estimate M by a process we call *skeletonizing*. Consider the isoprobability contours of an n-dimensional density. The k-skeleton is the best k-dimensional summary, an isoprobability contour (isopleth). To make this concrete, think in terms of the two-dimensional case. For linearly related X and Y with Gaussian distribution, the two-dimensional distribution will have elliptical isopleths, the optimal linear fit will be the major axis of the isopleths. The general strategy is then to estimate the k-skeleton for $k = 0,1,2,\ldots n$ and find the minimum k for which increasing the k does not significantly improve the fit.

We shall not develop the procedure in more detail in the present paper except to say that there are two key elements. One is the efficient estimation of high dimensional densities and the second is an efficient, consistent method for skeletonizing the isopleths. The parallel coordinate representation can, however, help in visualizing the process.

An important insight into the use of parallel coordinate diagrams is to recognize that structures map into structures (e.g. conics into conics) rather than just points into lines. One must, so to speak, see the forest rather than just the trees or more apropos, see the structure rather than just the lines. Recognizing dimensionality reduction is thus important. The process of searching for this structure involves an interactive process of transformation of coordinate axes (change of variables) and nonlinearly transforming individual axes until the template in Figure 3.2 is achieved.

Asimov (1985) suggests the idea of the grand tour as a method for searching for structure in Cartesian diagrams. The grand tour procedure consists of seeking a (reasonably) dense set of two-planes in n-dimensional space and sequentially projecting the data onto those two-planes. If the path through the set of two-planes is smooth, the visual appearance on a graphic screen will be a smooth flight through and around the data. Obviously this notion can be generalized to seeking a dense set of parallel-axes transformation.

This general framework for structural inference obviously still needs to be fleshed out. However, we include this discussion here to indicate that the parallel coordinate representation should be thought of as an element of a larger methodology and not merely as a simple n-dimensional graphical display device.

ACKNOWLEDGEMENTS

The data upon which Figures 4.1 and 4.2 are based were supplied by Dr. Fred Machell of the University of Texas Applied Research Laboratory. We are most pleased to acknowledge his help and advice. This research was supported by the Air Force Office of Scientific Research under Grant AFOSR-87-0179.

REFERENCES

1. Asimov, D. (1985), "The Grand Tour: A tool for viewing multidimensional data," *SIAM J. Scient. Statist. Comput.*, **6**, 128-143.

2. Baker, C.R. (Ed.), *Stochastic Processes in Underwater Acoustics,* Berlin: Springer-Verlag.

3. Baker, C.R. and Gualtierotti, A.F. (1986), "Likelihood ratios and signal detection for non-Gaussian processes," in *Stochastic Processes in Underwater Acoustics,* (C.R. Baker, Ed.), Berlin: Springer-Verlag, 154-180.

4. Carr, D.B., Nicholson, W.L., Littlefield, R. and Hall, D.L. (1986), "Interactive color display methods for multivariate data," in *Statistical Image Processing and Graphics,* (E. Wegman and D. DePriest, Eds.), New York: Marcel-Dekker, Inc., 215-250.

5. Chernoff, H. (1973), "Using faces to represent points in k-dimensional space," *J. Am. Statist. Assoc.*, **68**, 361-368.

6. Cleveland, W.S. and McGill, R. (1984), "The many faces of the scatterplot," *J. Am. Statist. Assoc.*, **79**, 807-822.

7. Dimsdale, B. (1984), "Conic transformations and projectivities," IBM Los Angeles Scientific Center Report #6320-2753.

8. Fienberg, S. (1979), "Graphical methods in statistics," *Am. Statist.*, **33**, 165-178.

9. Griffen, H.D. (1958), "Graphic computation of tau as a coefficient of disarray," *J. Am. Statist. Assoc.*, **53**, 441-447.

10. Inselberg, A. (1985), "The plane with parallel coordinates," *The Visual Computer,* **1**, 69-91.

11. Machell, F.W. and Penrod, C.S. (1984), "Probability density functions in ocean acoustic noise processes," in *Statistical Signal Processing,* (E. Wegman and J. Smith, Eds.), New York: Marcel-Dekker, Inc., 211-221.

12. Wegman, E.J. and Smith, J.G. (Eds.), (1984), *Statistical Signal Processing,* New York: Marcel-Dekker, Inc.

13. Wegman, E.J. (1986), "Hyperdimensional data analysis using parallel coordinates," Technical Report No. 1, Center for Computational Statistics and Probability, George Mason University, Fairfax, VA.

14. Wilson, G.R. and Powell, D.R. (1984), "Experimental and modeled density estimates of underwater acoustic returns," in *Statistical Signal Processing*, (E. Wegman and J. Smith, Eds.), New York: Marcel-Dekker, Inc., 223-239.

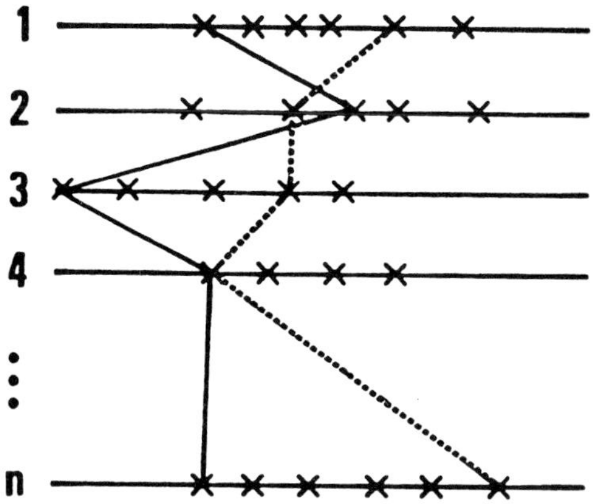

Figure 2.1 Parallel coordinate representation of two n-dimensional points.

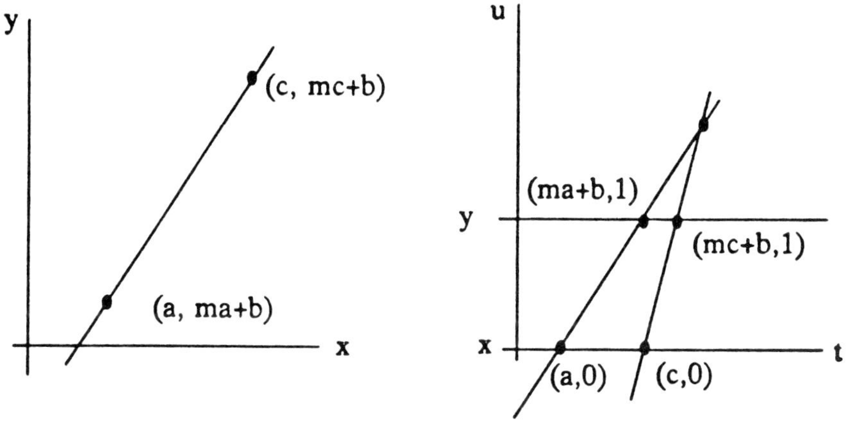

Figure 2.2 Cartesian and parallel coordinate plots of two points. The tu Cartesian coordinate system is superimposed on the xy parallel coordinate system.

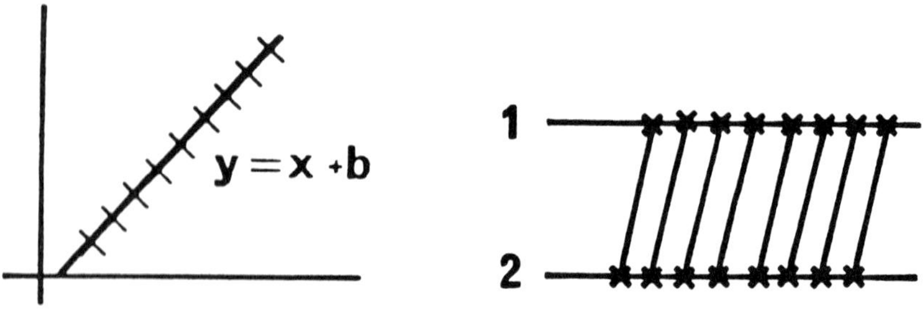

Figure 2.3a Cartesian and parallel coordinate plots of a line with slope 1.

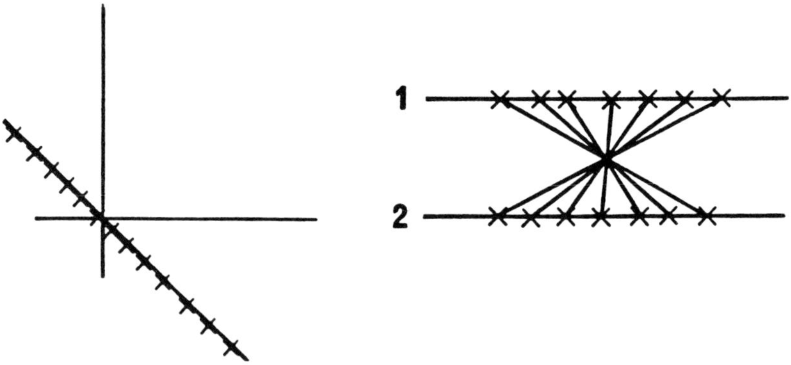

Figure 2.3b Cartesian and parallel coordinate plots of a line with slope -1.

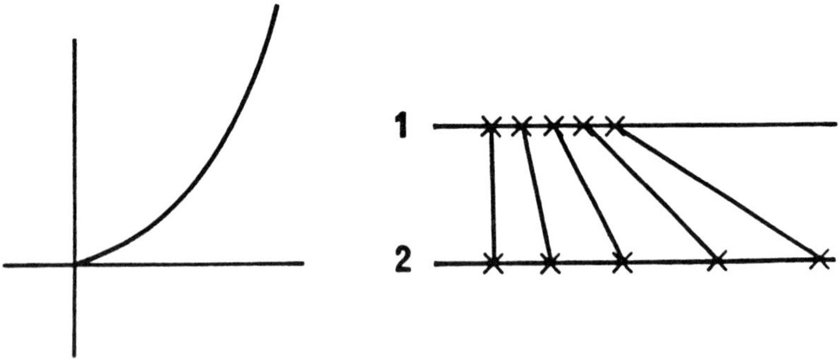

Figure 2.3c Cartesian and parallel coordinate plots of a nonlinear function.

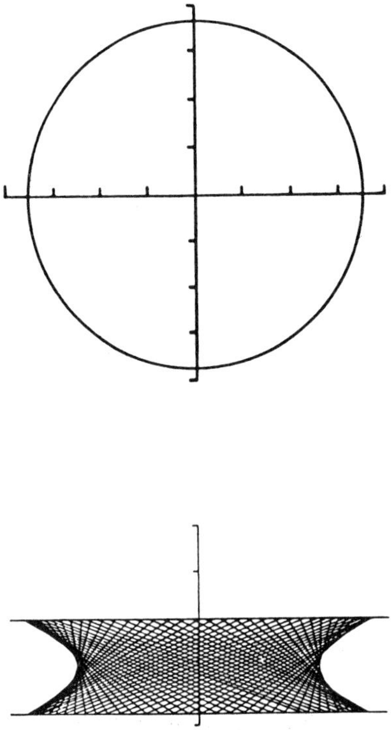

Figure 2.4 An ellipse in Cartesian coordinates is mapped into a line hyperbola in parallel coordinates with envelope a point hyperbola.

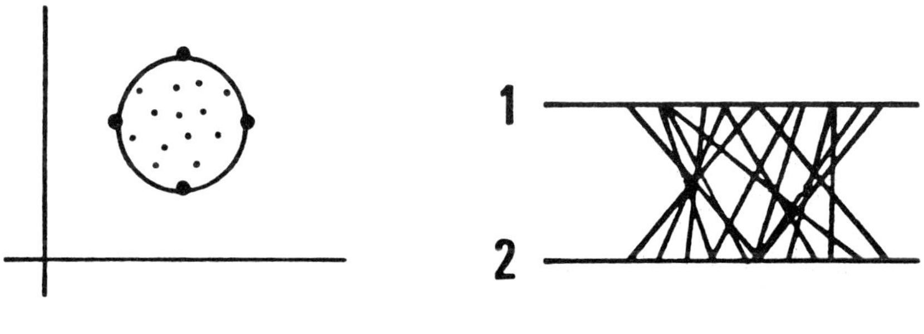

Figure 3.1a Uncorrelated data in a Cartesian scatter diagram would tend to an approximately circular convex ball, thus would tend to an approximately hyperbolic parallel coordinate diagram.

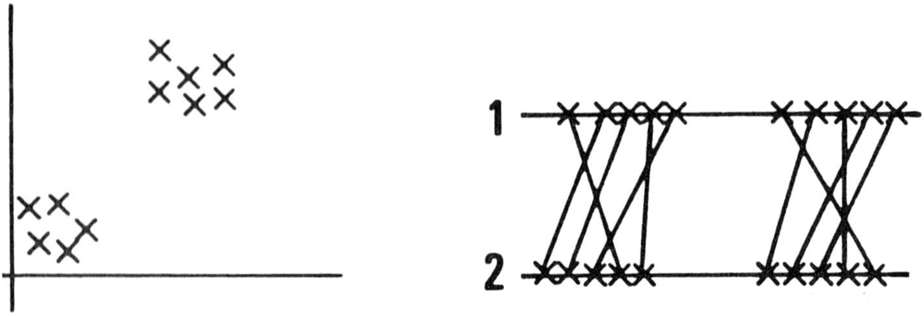

Figure 3.1b Clustered data separated both in x and y.

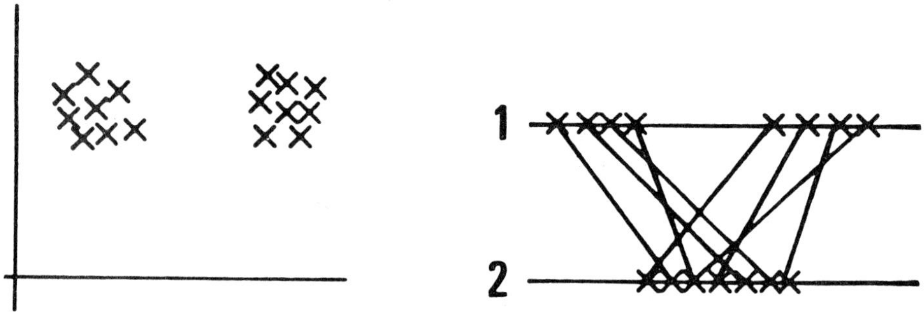

Figure 3.1c Clustered data separated in x but not in y. Note the diagnostic template, which is separation on any of the parallel coordinate axes, indicates clustering.

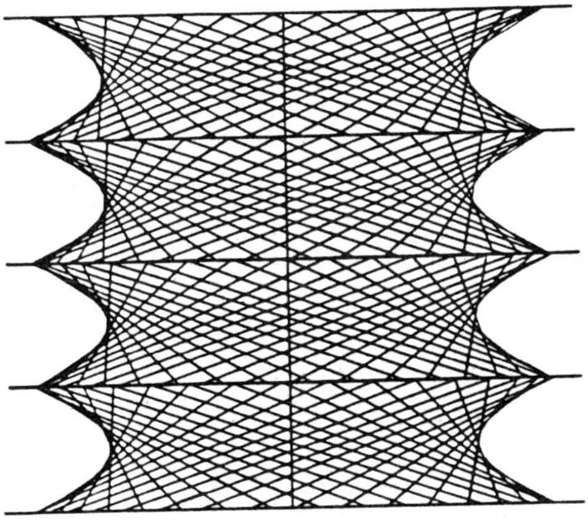

Figure 3.2 A five dimensional hypersphere in parallel coordinates.

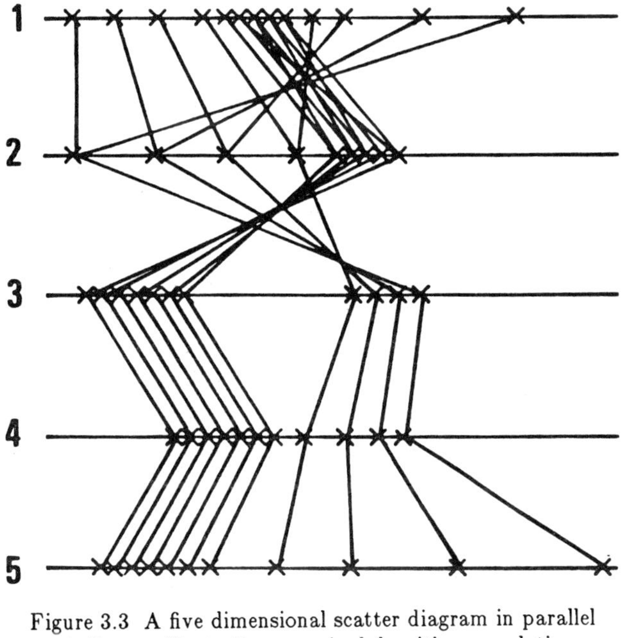

Figure 3.3 A five dimensional scatter diagram in parallel coordinates illustrating marginal densities, correlation, three dimensional clustering and a five dimensional mode.

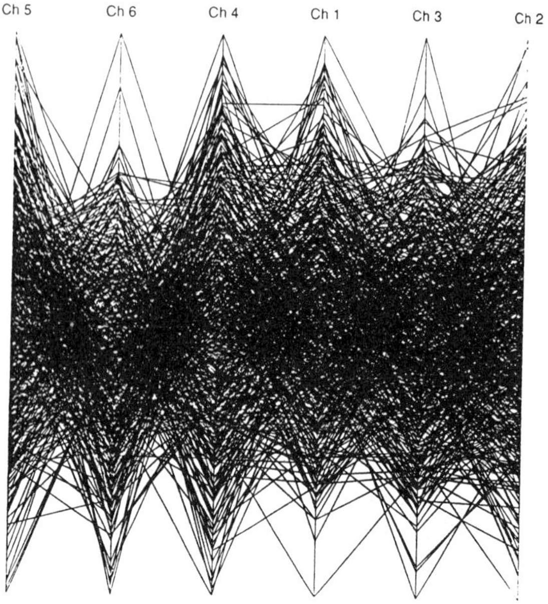

Figure 4.1 Parallel coordinate plot of six channel ambient ocean acoustic noise.

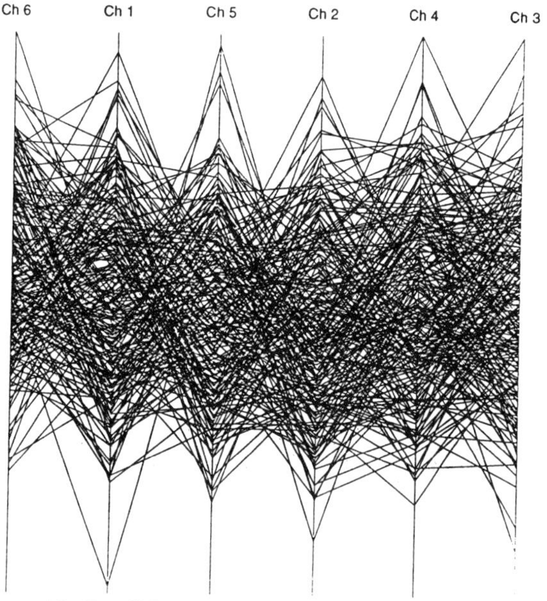

Figure 4.2 Parallel coordinate plot of six channel acoustic noise with broad band signal. The channels have been logged to show negative correlations.

PART II:

FILTERING, ESTIMATION AND REGRESSION

COMMENTS ON STRUCTURE AND ESTIMATION FOR NONGAUSSIAN LINEAR PROCESSES

by

M. Rosenblatt*
University of California, San Diego
La Jolla, California 92093

I. INTRODUCTION

A great deal of the research in time series analysis has been based on insights derived from the structure of Gaussian stationary processes. Suppose $\{X_n\}$, $n = \ldots,-1,0,1,\ldots$ is a Gaussian stationary sequence with mean zero and spectral distribution function $F(\lambda)$. If $F(\lambda)$ is absolutely continuous with derivative $f(\lambda) = F'(\lambda)$, the spectral density function, it follows that X_n has the representation

$$X_n = \sum_{k=-\infty}^{\infty} a_k V_{n-k} \tag{1}$$

with $\{V_n\}$ a sequence of Gaussian mean zero variance one random variables and a_k a sequence of real constants with

$$\sum_{k=-\infty}^{\infty} a_k^2 < \infty$$

where the derived Fourier series

$$a(e^{-i\lambda}) = \sum_{k=-\infty}^{\infty} a_k e^{-ik\lambda}$$

has the property that

$$\frac{1}{2\pi} | a(e^{-i\lambda}) |^2 = f(\lambda) . \tag{2}$$

There are an infinite number of functions $a(e^{-i\lambda})$ having the property (2) and one cannot distinguish them in terms of observations on the Gaussian process $\{X_n\}$. It has been the convention to resolve the question of the nonidentifiability of $a(e^{-i\lambda})$ in the Gaussian case if $\log f(\lambda)$ is integrable by taking it to be

$$c(e^{-i\lambda}) = \sum_{j=0}^{\infty} c_j e^{-ij\lambda} \tag{3}$$

where

$$c(z) = \sum_{j=0}^{\infty} c_j z^j = \sqrt{2\pi} \exp \left\{ \frac{1}{4\pi} \int_{-\pi}^{\pi} \log f(\lambda) \frac{e^{-i\lambda}+z}{e^{-i\lambda}-z} d\lambda \right\} \tag{4}$$

for $|z| < 1$. Among all functions $c(z)$ analytic in the unit disc with $\sum_{j=0}^{\infty} |c_j|^2 < \infty$ and

*Research supported in part by Office of Naval Research Contract N00014-81-K-003 and National Science Foundation Grant DMS 83-12106.

$$|c(e^{-i\lambda})|^2 = 2\pi f(\lambda),$$

the function (4) has maximal coefficient $c(0) = c_0$. In fact, it corresponds to the prediction problem for the process $\{X_n\}$ in a sense that we shall indicate. The stationary process $\{X_n\}$ has the Fourier representation

$$X_n = \int_{-\pi}^{\pi} e^{in\lambda}\, dZ(\lambda)$$

in terms of a process $Z(\lambda)$ of orthogonal increments

$$E\ Z(\lambda) \equiv 0$$

$$E[dZ(\lambda)d\bar{Z}(\mu)] = \delta_{\lambda,\mu}\, f(\lambda)d\lambda .$$

The best predictor of X_0 in terms of the past $X_{-1}, X_{-2},\ldots$ (in the sense of minimal mean square error of prediction) is

$$\int_{-\pi}^{\pi} \left(\sum_{j=1}^{\infty} c_j e^{-ij\lambda} \right) \{c(e^{-i\lambda})\}^{-1}\, dZ(\lambda)$$

and the variance of the prediction error is

$$2\pi \exp \left\{ \frac{1}{2\pi} \int_{-\pi}^{\pi} \log f(\lambda) d\lambda \right\}$$

Here $c(z)$ is given by (3). The function (3) clearly has no zeros in the unit disc and this is sometimes referred to as a minimum phase condition. Notice that the best predictor of X_0 is linear in the past because the process $\{X_n\}$ is Gaussian.

It is of some interest to consider autoregressive moving average (ARMA) schemes

$$\sum_{j=0}^{p} \alpha_j X_{n-j} = \sum_{k=0}^{q} \beta_k V_{n-k} \tag{5}$$

with $\alpha_0, \beta_0 \neq 0$. Assume that the sequence $\{V_n\}$ is one of independent Gaussian mean zero, variance one random variables. Assume that the polynomials

$$\alpha(z) = \sum_{j=0}^{p} \alpha_j z^j$$

$$\beta(z) = \sum_{k=0}^{q} \beta_k z^k$$

have no zeros in common. A stationary solution $\{X_n\}$ of the system of equations exists then if and only if

$$\alpha(e^{-i\lambda}) \neq 0$$

for all real λ. The stationary solution is unique and has an absolutely continuous spectral distribution function with spectral density

$$f(\lambda) = \frac{1}{2\pi} \left| \frac{\beta(e^{-i\lambda})}{\alpha(e^{-i\lambda})} \right|^2 .$$

One can consider the system (5) as giving one a recursive scheme that allows one to compute X_n given knowledge of $X_{n-1},\ldots,X_{n-p}$ and V_n, $V_{n-1},\ldots,V_{n-q}$. Given any reasonable initial conditions X_0, $X_1,\ldots,X_{p-1}$ the system (5) will be stable (in the sense that the computed X_n's do not diverge as $n \to \infty$) if and only if all the zeros of $\alpha(z)$ are outside $|z| \leq 1$. One can show that

$$\beta(e^{-i\lambda}) \,/\, \alpha(e^{-i\lambda})$$

is the function (3) corresponding to the prediction problem if and only if both $\beta(z)$ and $\alpha(z)$ have all their zeros outside $|z| \leq 1$.

The class of Gaussian stationary sequences (1) with absolutely continuous spectral distribution function is a small subset of the collection of linear processes

$$Y_n = \sum_{k=-\infty}^{\infty} a_k V_{n-k}, \quad \sum_{k=-\infty}^{\infty} a_k^2 < \infty \tag{6}$$

with $\{V_n\}$ a sequence of independent identically distributed random variables with mean zero and variance one. For a large class of nonGaussian linear processes we shall see that one can determine much more about the character of the function $a(e^{-i\lambda})$ than one can in the case of a Gaussian linear process from observations on the Y_n sequence alone. Let us again remark that in the Gaussian case only the modulus $|a(e^{-i\lambda})|$ of $a(e^{-i\lambda})$ can be resolved from observations on the Y_n sequence. Under appropriate conditions one can show that in the nonGaussian case $a(e^{-i\lambda})$ can be determined from observations on $\{Y_n\}$ up to sign and a factor $\exp\{in\lambda\}$ with n integral.

It is worthwhile making a few remarks on the prediction problem for a stationary nonGaussian ARMA process, that is, a stationary solution of the system of equations

$$\sum_{j=0}^{p} \alpha_j Y_{n-j} = \sum_{k=0}^{q} \beta_k V_{n-k}$$

with $\alpha_0, \beta_0 \neq 0$ and V_n nonGaussian. There is a stationary solution if and only if $\alpha(e^{-i\lambda}) \neq 0$ for all real λ and it is unique. The best predictor of Y_0 given the past $Y_{-1}, Y_{-2},\ldots$ in the sense of minimal mean square error of prediction is the linear predictor (obtained in the Gaussian case) mentioned above if and only if $\alpha(z), \beta(z)$ have no zeros in $|z| \leq 1$. If there are some zeros of $\alpha(z)$ or $\beta(z)$ in $|z| < 1$ the best predictor will be a nonlinear form in terms of the past.

II. ESTIMATION OF THE FUNCTION $a(e^{-i\lambda})$

Consider a linear process (6) with the property that the random variables V_n have finite moments up to order $k(>2)$. Let the corresponding cumulants of the random variables V_n be $\gamma_1, \ldots, \gamma_k$. The joint cumulant of $Y_n, Y_{n+\alpha_1},\ldots,Y_{n+\alpha_{s-1}}$, $s \leq k$, is

$$cum\,(Y_n, Y_{n+\alpha_1},\ldots,Y_{n+\alpha_{s-1}}) = \sum_k a_k\, a_{k+\alpha_1} \cdots a_{k+\alpha_{s-1}} \gamma_s \tag{7}$$

Let

$$b_s(\lambda_1, \ldots, \lambda_{s-1}) = (2\pi)^{-k+1} \sum_{\alpha_1,\ldots,\alpha_{s-1}} cum\,(Y_n, Y_{n+\alpha_1},\ldots,Y_{n+\alpha_{s-1}}) \exp\left(-\sum_{u=1}^{s-1} i\alpha_u \lambda_u\right)$$

be the sth order cumulant spectral density function of the process $\{Y_n\}$. Equation (7) implies that

$$b_s(\lambda_1,\ldots,\lambda_{s-1}) = \gamma_s a(e^{-i\lambda_1}) \cdots a(e^{-i\lambda_{s-1}}) \\ a(e^{i(\lambda_1+\cdots+\lambda_{s-1})}) . \tag{8}$$

At this point it is of some interest to give an example of a process that is not linear and that has a bispectral density of the form

$$b_3(\lambda_1,\lambda_2) = \gamma_3\, a(e^{-i\lambda_1})a(e^{-i\lambda_2})a(e^{i(\lambda_1+\lambda_2)}) . \tag{9}$$

Let $\{W_n\}$ be a stationary sequence that is a martingale difference sequence in both the forward and backward directions and yet is not a sequence of independent and identically distributed random variables. We shall explain what we mean by this remark. Let $\mathbf{B}_n = \mathbf{B}(W_j, j \le n)$ and $\mathbf{F}_n = \mathbf{B}(W_j, j \ge n)$ be the σ-fields generated by W_j, $j \le n$, and W_j, $j \ge n$ respectively. $\{W_n\}$ is a martingale difference sequence in the forward direction if

$$E\{W_n \mid \mathbf{B}_{n-1}\} \equiv 0$$

and a martingale difference sequence in the backward direction if

$$E\{W_n \mid \mathbf{F}_{n+1}\} = 0 .$$

Assume that $E\{W_n^3\} = \gamma_3 \ne 0$. Let

$$Y_n = \sum_{j=-\infty}^{\infty} a_{n-j} W_j , \quad \sum a_n^2 < \infty .$$

Now if $\{W_n\}$ is a martingale difference in both the forward and backward direction it follows that

$$E\{W_j W_k W_l\} = 0$$

if $j \le k \le l$ but not $j = k = l$. This immediately implies that (9) is valid. A simple example of a martingale difference in both forward and backward direction is constructed now. Let $\{\xi_j\}$ be a sequence of independent identically distributed random variables with mean zero $E\xi_j \equiv 0$ and $E|\xi_j|^3 < \infty$. Consider $f(\xi_{j+1},\ldots,\xi_{j+u})$, $u \ge 1$ a function of the u random variables $\xi_{j+1},\ldots,\xi_{j+u}$ with finite third moment. Set

$$W_j = \xi_j\, f(\xi_{j+1},\ldots,\xi_{j+u})\xi_{j+u+1} . \tag{10}$$

The sequence $\{W_j\}$ given by (10) is a martingale difference in both forward and backward directions because

$$E\{W_j \mid \ldots,\xi_{j+u-1},\xi_{j+u}\} = 0$$

$$E\{W_j \mid \xi_{j+1},\xi_{j+2},\ldots\}.$$

We now consider how information about the transfer function $a(e^{-i\lambda})$ can be recovered in the case of a nonGaussian linear process.

Proposition. *Assume that $\{Y_n\}$ is a nonGaussian linear process with*

$$\sum_j |j|\,|a_j| < \infty$$

and $a(e^{-i\lambda}) \ne 0$ for all real λ. Further, let the independent, identically distributed random variables V_n have some cumulant of order $s > 2$, $\gamma_s \ne 0$. Then the function $a(e^{-i\lambda})$ can be identified in terms of observations on $\{Y_n\}$ alone up to an integer n in a factor $e^{in\lambda}$ and the sign of $a(1) = \sum_k a_k$.

The cumulant spectral density of order s is given by (8). Let

$$h(\lambda) = arg \left\{ a(e^{-i\lambda}) \frac{a(1)}{|a(1)|} \right\}.$$

We then have

$$\left\{ \frac{a(1)}{|a(1)|} \right\}^s \gamma_s = (2\pi)^{s/2-1} b_s(0,\ldots,0)/\{f(0)\}^{s/2}$$

and

$$h(\lambda_1) + \cdots + h(\lambda_{s-1}) - h(\lambda_1 + \cdots + \lambda_{s-1})$$
$$= arg \left[\left\{ \frac{a(1)}{|a(1)|} \right\}^s \gamma_s^{-1} b_s(\lambda_1, \ldots, \lambda_{s-1}) \right]$$

because $h(-\lambda) = -h(\lambda)$. Also

$$h'(0) - h'(\lambda) = \lim \frac{1}{(s-2)\Delta} \{h(\lambda) + (s-2)h(\Delta) - h(\lambda + (s-2)\Delta)\} .$$

Further

$$h(\lambda) = \int_0^\lambda \{h'(u) - h'(0)\} du + c\lambda = h_1(\lambda) + c\lambda$$

with $c = h'(0)$. Thus

$$h(\pi) = h_1(\pi) + c\pi .$$

The fact that the a_j's are real implies that $h(\pi) = a\pi$ for some integer a. If $h(\pi)/\pi = \delta$, $h(\pi) = a\pi = (\delta + c)\pi$ and so $c = a - \delta$. The integer a cannot be specified without additional information since a change in a corresponds to reindexing of the $\{V_n\}$ sequence. The sign of $a(1)$ cannot be determined since one can multiply the a_j's and V_n's by (-1) without changing the process $\{Y_n\}$. Since

$$a(e^{-i\lambda}) = \{2\pi f(\lambda)\}^{1/2} \exp\{ih(\lambda)\}$$

the Proposition follows. This proposition was essentially initially given in Rosenblatt (1980).

It is clear that one can estimate $f(\lambda)$ by a standard spectral density estimator. We shall assume that $\gamma_3 \neq 0$ and show how to get phase information or estimate $h(\lambda)$ under this assumption. This would be a plausible assumption if the data appears to have a decidedly asymmetric marginal distribution about zero. One would make use of relation (9). If the data appeared to have a symmetric marginal distribution, the procedure based on third order cumulant spectra (or bispectra) would be inappropriate. A procedure based on $\gamma_4 \neq 0$ and corresponding use of fourth order cumulant spectra (using relation (8) for $s = 4$) would then be appropriate. The fourth order procedure has many points of similarity with the third order procedure and is described in Lii and Rosenblatt (1985) in detail. For that reason we shall only describe the third order method in detail.

Notice that in the case $s = 3$

$$h'(0) - h'(\lambda) = \lim_{\Delta \to 0} \frac{1}{\Delta} \{h(\lambda) + h(\Delta) - h(\lambda + \Delta)\}$$

and up to an ambiguity in the sign of $b_3(0,0)$ one has

$$h(\lambda) + h(\Delta) - h(\lambda + \Delta) = arg\ \{b_3(\lambda,\Delta)\}\ .$$

At this point we shall drop the subscript and take for granted that we are dealing with the bispectral density $b(\lambda,\mu)$. Assume that ${}_n b(\lambda,\mu)$ is a consistent estimate of $b(\lambda,\mu)$ (as $n \to \infty$) based on a sample of size n. A plausible estimate of

$$\theta(\lambda,\mu) = arg\ b(\lambda,\mu)$$

is given by

$$\theta_n(\lambda,\mu) = arctan\ (Im\ {}_n b(\lambda,\mu)/\mathrm{Re}\ {}_n b(\lambda,\mu))\ .$$

One can then show that

$$\begin{aligned}\theta_n(\lambda,\mu) - \theta(\lambda,\mu) = &- \frac{\mathrm{Im}\ b(\lambda,\mu)}{|b(\lambda,\mu)|^2}\{\mathrm{Re}\ {}_n b(\lambda,\mu) - \mathrm{Re}\ b(\lambda,\mu)\} \\ &+ \frac{\mathrm{Re}\ b(\lambda,\mu)}{|b(\lambda,\mu)|^2}\{\mathrm{Im}\ {}_n b(\lambda,\mu) - \mathrm{Im}\ b(\lambda,\mu)\} \\ &+ o({}_n b(\lambda,\mu) - b(\lambda,\mu))\ .\end{aligned} \tag{11}$$

Let

$$z = x + iy = re^{i\theta}$$

with $r = |z|$ and a principal value determination of $\theta = arctan\ (y/x)$. Then

$$\frac{\partial \theta}{\partial y} = \frac{\theta}{r^2},\ \frac{\partial \theta}{\partial x} = -\ \frac{y}{r^2},\ \frac{\partial^2 \theta}{\partial x^2} = \frac{2xy}{r^4}\ ,$$

$$\frac{\partial^2 \theta}{\partial y^2} = -\ \frac{2xy}{r^4}\ ,\ \frac{\partial^2 \theta}{\partial x \partial y} = \frac{1}{r^2} - \frac{2x^2}{r^4}$$

and (11) follows immediately on using these relations in a Taylor expansion of the arctan function.

Now our principal object is to estimate

$$h_1(\lambda) - \frac{h_1(\pi)}{\pi}\lambda\ .$$

Suppose we set $\Delta = \Delta(n)$, $k\Delta = \lambda$ and let $\Delta = \Delta(n) \to 0$ as $n \to \infty$. For convenience assume that $b(0,0)$ is positive. Later by making use of a simple modification it will be seen how to take care of the case in which $b(0,0)$ is negative. Notice that

$$\begin{aligned} h_1(\lambda) = h(\lambda) - h'(0)\lambda &\cong h(k\Delta) - \frac{h(\Delta)}{\Delta}k\Delta \\ &= -\sum_{j=1}^{k-1}\{h(j\Delta) + h(\Delta) - h((j+1)\Delta)\} \\ &= -\sum_{j=1}^{k-1} arg\ b(j\Delta,\Delta)\ . \end{aligned}$$

This suggests that

$$H_n(\lambda) = -\sum_{j=1}^{k-1} arg\ {}_n b(j\Delta,\Delta)$$

would be a plausible estimate of $h_1(\lambda)$.

Let $EY_n^6 < \infty$. Also assume that $a(1) > 0$. The case in which $a(1) < 0$ can be taken care of by a simple modification of our discussion. Consider $H_n(\lambda)$, $k\Delta = \lambda$, as an estimate of $h_1(\lambda)$ with the bispectral estimates ${}_nb(j\Delta,\Delta)$ weighted averages of third order periodogram values (see Rosenblatt (1983)). If the bispectral density is twice continuously differentiable, the weight function is symmetric and bandlimited with bandwidth Δ, then

$$H_n(\lambda) - h_1(\lambda) = R_n(\lambda) + o(H_n(\lambda) - h_1(\lambda)) . \tag{12}$$

Further

$$ER_n(\lambda) \sim - \int_0^\lambda \tfrac{1}{2} \{b(u,0)\}^{-1} \{AD_u^2 \operatorname{Im} b(u,0) \tag{13}$$

$$+ 2BD_uD_v \operatorname{Im} b(u,0)$$

$$+ CD_v^2 \operatorname{Im} b(u,0)\} du \;\; \Delta + o(\Delta)$$

with A, B, and C the second moments of weight function w of the bispectral estimates

$$A = \int u^2 w(u,v) du dv , \qquad B = \int uvw(u,v) du dv ,$$

$$C = \int v^2 w(u,v) du dv .$$

Here D_u and D_v are the partial derivatives with respect to u and v. Also

$$cov \; \{R_n(\lambda), R_n(\mu)\} \sim \frac{2\pi^2}{\Delta^3 n \gamma_3^2} \min(\lambda,\mu) \int w^2(u,v) du dv . \tag{14}$$

It is understood here that $\Delta = \Delta(n) \to 0$, $\Delta^2 n \to \infty$ as $n \to \infty$. A derivation of these estimates is given in Lii and Rosenblatt (1982). The results on asymptotic behavior of bias and variance of estimates of $H_n(\lambda)$ hold also for moving averages of a backward and forward martingale sequence. The deconvolution procedure can be applied to such sequences.

The sign of $a(1)$ may be negative. The sign of $a(1)$ can be estimated by observing the real part of ${}_nb(\Delta,\Delta)$. If the sign is negative multiply all the ${}_nb(j\Delta,\Delta)$ by minus one. The estimate $H_n(\lambda)$ should then be given by

$$H_n(\lambda) = -\sum_{j=1}^{k-1} arg \; \{- {}_nb(j\Delta,\Delta)\} .$$

Assume that one has a sample of size $n = kN$. Center and normalize the data so that it has mean zero and variance one. The data should be broken up into k disjoint sections of length N so that the variance of the bispectral estimate of each section is not large. Determine a grid of points $\lambda_j = j\Delta$ in $(0,2\pi)$, $j = 1,...,M$, $\Delta = 2\pi L/N$. Compute the bispectral estimate ${}_Nb(j\Delta,\Delta)$ of the type mentioned with a weight function of bandwidth Δ from each section. The estimates from the different sections should be averaged so as to obtain a final estimate ${}_nb(j\Delta,\Delta)$. An extensive discussion of this type of algorithm is given in Lii and Helland (1981). An estimate of $a(e^{-i\lambda})$ up to the indeterminacy noted in the proposition can be formed in an obvious manner from $H_n(\lambda)$

and an estimate $f_n(\lambda)$ of the spectral density function. Notice that

$$a_k = \frac{1}{\pi} \int_0^{2\pi} a(e^{-i\lambda}) e^{ik\lambda} d\lambda .$$

This suggests an estimate $\hat{a}_k$ of a_k given by

$$\hat{a}_k = \frac{1}{2\pi} \int_0^{2\pi} \hat{a}(e^{-i\lambda}) e^{ik\lambda}\, d\lambda \tag{15}$$

$$\cong \frac{1}{(M+2)} \sum_{j=0}^{M+1} \sqrt{2\pi f_n(\lambda_j)} \exp\left\{ i(H_n(\lambda_j) - \frac{H_n(\pi)}{\pi} \lambda_j + k\lambda_j) \right\}.$$

Estimates of the Fourier coefficients of the function $a(e^{-i\lambda})^{-1}$ can be computed in an analogous manner. These are useful in deconvolution, that is, effecting estimates of the V_k's from knowledge of the Y_k's without prior knowledge of the filter transform $a(e^{-i\lambda})$. The problem of deconvolution is important in a class of geophysical problems (see Donoho (1981), Godfrey and Rocca (1981), and Wiggins (1978)). Estimates of $H_n(\lambda)$ are useful in modifying linear processes so that they are zero phase. Such adjustments are useful in some geophysical contexts.

The graph illustrates the application of the deconvolution procedure to a moving average

$$Y_t = V_t - 5V_{t-1} + 6.25V_{t-2} - 1.25V_{t-3} + 1.5V_{t-4}$$

of independent Pareto variables V_t with common density

$$f(v) = 4v^{-5},\ v \geq 1 .$$

The roots of the polynomial

$$1-5z + 6.25z^2 - 1.25z^3 + 1.5z^4$$

are 1/2, 1/3, $\pm 2i$. From this it is clear that the polynomial is not minimum phase. The sample size is 1280. The first line is a graph of the sequence Y_t. The second line graphs the generating V_t sequence. The third line gives the result of the deconvolution procedure applied to the sequence Y_t. The fourth line gives the error in the deconvolution, that is, the difference between lines two and three. The last line gives the result of a minimum phase deconvolution applied to Y_t.

It is of some interest to look at the deconvolution problem with the nonGaussian data $\{Y_n\}$ perturbed by a small Gaussian noise sequence $\{\eta_n\}$ independent of $\{Y_n\}$. The observations $W_1,\ldots,W_n$ are now given by

$$W_k = Y_k + \eta_k$$

with the Y_k sequence a nonGaussian linear process. The weights a_k of the linear scheme are assumed to be unknown. Let us see what happens in the deconvolution procedure described earlier in this circumstance. The object is to obtain estimates of the V_k's from observations on the W_k's. The bispectral density of the W sequence is the same as that of the Y sequence because $\{\eta_k\}$ is Gaussian and independent of the Y sequence. This suggests that if we compute $H_n(\lambda)$ as before, its asymptotic properties as given by (12), (13) will be the same. But (14) is modified with the factor $\min(\lambda,\mu)$ replaced by

$$\int_0^{\min(\lambda,\mu)} \left(1 + \frac{f_{\eta\eta}(0)}{f(0)}\right) \left(l + \frac{f_{\eta\eta}(u)}{f(u)}\right)^2 du$$

with $f_{\eta\eta}$ the spectral density of the η process. However, in making an estimate of $a(e^{-i\lambda})$, the coefficients a_k (see (15)), or the coefficients of the deconvolution filter, an estimate of the spectral density is employed. Let ${}_n f_{WW}(\lambda)$ be a standard estimate of the spectral density of the W process. Assume that one uses $|{}_n f_{WW}(\lambda)|^{1/2}$ in (15), for example, instead of $|{}_n f_{yy}(\lambda)|^{1/2}$ which is no longer available. The spectral density of the W process is $f_{yy}(\lambda) + f_{\eta\eta}(\lambda)$. Now

$$|{}_n f_{WW}(\lambda)|^{1/2} = |f_{WW}(\lambda)|^{1/2} |1 + \frac{{}_n f_{WW}(\lambda) - f_{WW}(\lambda)}{f_{WW}(\lambda)}|^{1/2}$$

$$= |f_{WW}(\lambda)|^{1/2} \left\{ 1 + \frac{1}{2} \frac{{}_n f_{WW}(\lambda) - f_{WW}(\lambda)}{f_{WW}(\lambda)} (1 + o(1)) \right\}.$$

The factor

$$|f_{WW}(\lambda)|^{1/2} = |f_{YY}(\lambda)|^{1/2} \left\{ 1 + \frac{1}{2} \frac{f_{\eta\eta}(\lambda)}{f_{YY}(\lambda)} (1 + o(1)) \right\}.$$

If $f_{\eta\eta}(\lambda)/f_{YY}(\lambda)$ is small and n is sufficiently large we would not expect the estimate to be perturbed to an appreciable degree by the noise η. It is clear that the intermediate estimates given here could be put together into final estimates of error in a procedure like deconvolution.

REFERENCES

1. Donoho, D. (1981). "On minimum entropy deconvolution," in *Applied Time Series II* (ed. D. F. Findley) 565-608.

2. Godfrey, R., and Rocca, F. (1981), "Zero memory nonlinear deconvolution," Geophysical Prospecting 29, 189-228.

3. Lii, K. S., and Helland, K. N. (1980), "Cross-bispectrum computation and variance estimation," ACM Trans. Math. Software 7, 284-294.

4. Lii, K. S., and Rosenblatt, M. (1982), "Deconvolution and estimation of transfer function phase and coefficients for nonGaussian linear processes," Ann. Statist. 10, 1195-1208.

5. Lii, K. S., and Rosenblatt, M. (1984), "Non-Gaussian linear processes, phase and deconvolution" in *Statistical Signal Processing* (ed. E. J. Wegman and J. G. Smith) 51-58.

6. Lii, K. S., and Rosenblatt, M. (1985), "A fourth order deconvolution technique for nonGaussian linear processes," to appear in Multivariate Analysis VI.

7. Rosenblatt, M. (1980), "Linear processes and bispectra," J. Appl. Prob. 17, 265-270.

8. Rosenblatt, M. (1983), "Cumulants and cumulant spectra," in *Time Series in the Frequency Domain* (ed. D. Brillinger and P. Krishnaiah) 369-382, North Holland.

9. Wiggins, R. A. (1978), "Minimum entropy deconvolution," Geoexploration **17**.

HARMONIZABLE SIGNAL EXTRACTION, FILTERING AND SAMPLING

M.M. Rao
University of California
Riverside, California 92521

I. INTRODUCTION

The purpose of this paper is to describe some aspects of analysis on a class of nonstationary and non Gaussian processes dealing with linear filtering, signal extraction from observed data, and sampling the process. The class to be considered consists of harmonizable processes which uses some suitably generalized spectral methods of the classical theory. Let us elaborate these statements.

If $\{X_t, t \in T\}$ is a second order process with zero means, on a probability space, usually written $X_t \in L_0^2(P)$ for all $t \in T$, let r be its covariance function. Then the process (= time series) is *stationary* whenever r admits a representation

$$E(X_s \bar{X}_t) = r(s,t) = \int_{\hat{T}} e^{i(s-t)\lambda} F(d\lambda), \qquad s,t \in T, \tag{1}$$

where $\hat{T} = (-\pi, \pi]$, if T = the integers $\mathbb{Z}$, and $\hat{T} = \mathbb{R}$, if T = the reals $\mathbb{R}$. *These are the only two types considered;* the first one is termed a *discrete* and the second a *continuous* parameter process. Here F is a nonnegative, nondecreasing bounded function, and the integral is in the standard Lebesgue sense. Processes which are not stationary but which are subject to such a Fourier analysis can be used for many purposes for which stationary ones have been used even when that assumption need not be true. The class generalizing the latter is called the harmonizable family and it can be introduced at two levels of generality. Thus one of these is termed *strongly* (or Loève) *harmonizable* if the covariance function r can be expressed as

$$r(s,t) = \int_{\hat{T}} \int_{\hat{T}} e^{is\lambda - it\lambda'} F(d\lambda, d\lambda'), \qquad s,t \in T, \tag{2}$$

where $\hat{T}$ is as before and $F: \hat{T} \times \hat{T} \to \mathbb{C}$ is a complex valued positive definite function of bounded variation, i.e., F is again a covariance function of bounded variation on $\hat{T} \times \hat{T}$. The integral in (2) is also in the Lebesgue sense. This generalization is due to Loève [18]. If F concentrates on $\lambda = \lambda'$ of $\hat{T} \times \hat{T}$, then (2) reduces to (1), and stationarity is included in this extension. The following simple example describes a strongly harmonizable process which is not stationary. Let $f: \mathbb{R} \to \mathbb{R}$ be a (Lebesgue) integrable function with $\hat{f}$ as its Fourier transform, i.e.,

$$\hat{f}(t) = \int_{\mathbb{R}} e^{itx} f(x) dx, \qquad t \in \mathbb{R}.$$

Let $X_t = \hat{f}(t)\xi$ where ξ is a random variable with mean zero and unit variance. Then $r(s,t) = \hat{f}(s)\bar{\hat{f}}(t)$, and it admits a representation (2) with $F(d\lambda, d\lambda') = f(\lambda)\bar{f}(\lambda')\, d\lambda d\lambda'$. Since this F does not concentrate on $\lambda = \lambda'$, the process is strongly harmonizable but not

This research is supported, in part, under an ONR Contract.

stationary.

Consider, on the other hand, a stationary process $\{X_t, t\in T\}$ and let τ be a truncation operator on such processes, i.e. $\tau X_t = X_t$, if $t \geq a$, $= 0$ if $t < a$, then $\{\tau X_t, t\in T\}$ need not be stationary and can also be not strongly harmonizable. More generally, if V is a bounded linear operator on $L_0^2(P)$, and $\{X_t, t\in T\}$ is strongly harmonizable, then $\{VX_t, t\in T\}\subset L_0^2(P)$ is not necessarily strongly harmonizable. These statements are not obvious, but they will follow from the results given later on. Thus one needs a suitable extension of (2) in order to accommodate these cases.

A second order process $\{X_t, t\in T\} \subset L_0^2(P)$ with covariance r, is termed *weakly harmonizable* if r can be represented as (2) with F as a covariance function of only finite *Fréchet variation* so that

$$\|F\|(\hat{T}\times\hat{T}) = \sup\{ \sum_{i,j=1}^{n} a_i \bar{a}_j F(A_i, A_j): |a_i| \leq 1, a_i \in \mathbb{C}, A_i \subset \hat{T},$$
$$\text{disjoint Borel sets}, i=1,2,\ldots,n; n \geq 1\} < \infty.$$

In this case the integral of (2) is defined in the sense of Morse and Transue [20], and it is a nonabsolute integral (in some ways reminding the Riemann integral). It turns out that this class has the desired closure properties so that if $\{X_t, t\in T\}$ is either stationary, strongly or weakly harmonizable and V is any bounded linear operator on $L_0^2(P)$, then $Y_t = VX_t, t \in T$, is always weakly harmonizable, and this freedom is needed for the (linear) filtering problems in the subject. Consequently most of the work described below is in terms of these (weakly) harmonizable processes which, moreover, are never assumed Gaussian.

Closely related to the harmonizable family is a class of nonstationary time series introduced by Kampé de Feriet and Frenkiel, starting in the middle 1950's, and expounded in detail in [14], to be called *class* (KF). It is a second order process $\{X_t, t\in T\}$ as above, whose covariance function r satisfies a first arithmetical mean limit condition, i.e., the following limits should exist:

$$\lim_{n\to\infty} r_n(k) = \lim_{n\to\infty} \frac{1}{n} \sum_{j=0}^{n-|k|-1} r(j, j+|k|), \qquad k \in \mathbb{Z}, \tag{3}$$

or

$$\lim_{\alpha\to\infty} r_\alpha(h) = \lim_{\alpha\to\infty} \frac{1}{\alpha} \int_0^{\alpha-|h|} r(s, s+|h|)ds, \qquad h \in \mathbb{R}. \tag{3'}$$

It will be seen later that each strongly harmonizable process is in class (KF), but there exist weakly harmonizable processes which do not belong to this class. On the other hand it is easily shown that a Brownian motion process is in class (KF), but it is not harmonizable. Thus all these three (distinct) classes generalize stationarity, and can be used in modeling the physical phenomena which are nonstationary but not explosive.

The weakly harmonizable class was first introduced by Bochner [1] and [2], under the name "V-bounded process" and independently by Rozanov [31] (unfortunately) again under the name "harmonizable". Moreover, certain other related boundedness criteria were considered by Bochner in [2]. The signal extraction and filtering problems will be modeled with these nonstationary processes in view. The formulas with the least squares error criterion for optimality show in what sense some of the stationary formulas are "robust", i.e., have analogous functional forms. This would imply that a study of the harmonizable classes for applications is advantageous. Thus the filtering and sampling problems will be treated in this paper. They show new areas of interest for applications.

II. THE GENERAL LINEAR MODEL

One of the fundamental features of (both) harmonizable processes as well as the stationary ones is that such a process $\{X_t, t \in T\}$ admits an integral representation:

$$X_t = \int_{\hat{T}} e^{it\lambda} Z(d\lambda), \quad t \in T, \tag{4}$$

where $Z(\cdot)$ is a σ-additive (in the mean square sense) set function on the Borel sets of $\hat{T}$, $E(Z(A)) = 0$, $A \subset \hat{T}$, and for any pair of Borel sets A, B of $\hat{T}$, one has

$$E(Z(A)\overline{Z(B)}) = \begin{cases} F(A,B), \\ F(A \cap B), \end{cases} \tag{5}$$

accordingly as the process is harmonizable or stationary. The F in the harmonizable case is that given by (2) (with finite ordinary or Fréchet variations respectively), and by (1) in the stationary case. The stochastic integral in (4) is defined with one of the standard methods in Probability Theory [11] or it can be interpreted (as will be done in this paper) as a special (i.e., Hilbert space valued) Dunford-Schwartz integral ([10], IV. 10.7). In the time series analysis context, such a $Z(\cdot)$ in (4) is called a *stochastic spectral measure* of the process, and that of the covariance function r, given by (1) or (2), is termed a *spectral function.* Both are defined on $\hat{T}$, the *frequency (or spectral) domain* of the process, while T is called the *time domain.* The details of (4) may be found in [22] and [29].

Let $\{X_t, t \in T\}$ and $\{Y_t, t \in T\}$ be processes of the above types related by the equation

$$\Lambda X_t = Y_t, \quad t \in T, \tag{6}$$

where Λ is a linear operator on $L_0^2(P)$ into itself, such that it commutes with translations of t. The general form of Λ, to be considered here, is a difference-differential operator with *constant* coefficients, i.e., one of the form

$$\Lambda X_t = \sum_{j=0}^{l} \sum_{i=0}^{k} a_{ij} X^{(i)}(t - \tau_j) = Y_t, \quad t \in T, a_{ij} \in \mathbb{C} \tag{7}$$

Here $X^{(i)}$ is the ith-derivative in the mean- square sense. Thus if $k=1$, one has (X_t and $X(t)$ are the same)

$$E\left(\left|\frac{X_{t+h} - X_t}{h} - X'(t)\right|^2\right) \to 0, \text{as} \quad h \to 0,$$

and higher order derivatives are defined iteratively. The second kind of Λ is an integro-differential operator of the form

$$Y_t = \Lambda X_t = \sum_{i=0}^{k} \int_T X^{(i)}(t-\lambda) H_i(d\lambda), t \in T, \tag{8}$$

where each H_i is of (the usual) bounded variation on T. (Whenever the derivatives are admitted, one naturally takes $T = \mathbb{R}$.)

If the problem is (linear) filtering, then (6) is interpreted as follows. The operator Λ is known except for the coefficients a_i (or H_i) and the *output process* $\{Y_t, t \in T\}$ is at our disposal. If only the covariance structure of the X_t-process is assumed known (from prior considerations), it is desired to obtain the input process $\{X_t, t \in T\}$ from the output. Thus one should find (necessary and) sufficient conditions on the filter operator Λ, in order that the input can be obtained (by inversion) from the output. Suppose on the other hand that $\Lambda = I + \Lambda_1$, implying

$$Y_t = X_t + \Lambda_1 X_t = \tilde{X}_t + S_t \text{(say)}. \tag{9}$$

Then $\{S_t, t \in T\}$ may be regarded as an unknown *stochastic signal* and $\tilde{X}_t$ is the (unknown) noise whereas Y_t is the (output or) observed series. The problem here is to extract (or estimate) the signal from noise having only known their covariance structure and the output process. These are the models (7)-(9), which can be thought of as specializations of the general linear model (6). Several aspects of these are treated below in some detail.

Before entering into the analysis of processes satisfying (7)-(9), it is useful to investigate the role played by the stationarity in (6). Various forms of Λ in (6) have been considered in the literature, but primarily for the stationary class. The general linear operations on second order processes were first discussed by Karhunen [15], and the models (7) and (8) with $k=0$ were studied by Nagabhushanam [21]. The general forms were given by Bochner [1] and [2] who introduced the weakly harmonizable class for the first time in that work. His primary interest there was to study the solution process $\{X_t, t \in T\}$ if the output $\{Y_t, t \in T\}$ is stationary. His main conclusion is formulated below. Roughly, it says that for most linear operations Λ, if X_t is a solution of (6) which is "V-bounded" (i.e., weakly harmonizable), then it is possible to construct a stationary process X_t^o (in addition to the nonstationary solution) such that $\Lambda X_t^o = Y_t$, and it generalizes the work of [21] to a great extent. On the other hand, this need not obtain if the output process $\{Y_t, t \epsilon T\}$ itself is not stationary, as is the case in the present paper. Here Bochner's result will be given first, and then a classification of several second order processes admitting an integral representation (4) will be noted since that is useful for our analysis. Finally the related class (KF) is discussed and then filtering and sampling are also treated. A number of problems to be studied are pointed out along the way.

Consider the linear (filter) operator given by (7) or (8) and the processes representable by (4). Then on substitution one gets

$$\begin{aligned} \Lambda X_t &= \sum_{j'=0}^{l} \sum_{j=0}^{k} a_{j' j} X^{(j)}(t - u_{j'}) \\ &= \int_{\hat{T}} \Big(\sum_{j'=0}^{l} \sum_{j=0}^{k} a_{j' j} (i\lambda)^j \, e^{-iu_{j'}\lambda} \Big) e^{it\lambda} Z_x(d\lambda) \\ &= \int_{\hat{T}} \tilde{C}(\lambda) e^{it\lambda} Z_x(d\lambda), (say). \end{aligned} \tag{10}$$

The function $\tilde{C}$: $\lambda \mapsto \tilde{C}(\lambda)$ is the *spectral characteristic* of the difference-differential filter which governs the behavior of Λ, given by (7). A slightly more involved computation is needed for (8).

Thus, letting $D_t^j X_t = X_t^{(j)}$, one has for (8):

$$\Lambda X_t = \sum_{j=0}^{k} \int_T D_t^j \Big(\int_{\hat{T}} e^{i(t-u)\lambda} Z_x(d\lambda) \Big) H_j(du)$$

$$= \sum_{j=0}^{k} \int_T \int_{\hat{T}} (i\lambda)^j e^{i(t-u)\lambda} Z_x(d\lambda) H_j(du),$$

since D_t^j and the integral commute by ([10], IV. 10.8 (f)) which is valid for D_t^j also, (cf., [13], Thm. 3.7.12),

$$= \int_{\hat{T}} \Big[\sum_{j=0}^{k} (i\lambda)^j \int_T e^{-iu\lambda} H_j(du) e^{it\lambda} Z_x(d\lambda), \tag{11}$$

by Thm. 3.7.13 of [13], but an analog of Fubini's theorem can be easily obtained in this particular case,

$$= \int_{\hat{T}} C_1(\lambda) e^{it\lambda} Z_x(d\lambda), (say).$$

The function $C_1 : \lambda \mapsto C_1(\lambda)$ is called the *spectral characteristic* of the integro-differential operator (filter) Λ of (8), and its behavior for $|\lambda| \to 0$ and $|\lambda| \to \infty$ governs the filter of (8). An introduction to the applications of filters may be found in ([11], Section V.5).

If the output process is also representable by (4) with $Z_y(\cdot)$ as its stochastic spectral measure, then (6) becomes on using (10) and (11):

$$\int_{\hat{T}} C(\lambda) e^{it\lambda} Z_x(d\lambda) = \int_{\hat{T}} e^{it\lambda} Z_y(d\lambda), \qquad t \in T, \tag{12}$$

where C is $\tilde{C}$ or C_1 accordingly as (7) or (8) is considered. In case the Y_t-process is stationary, then $Z_x(\cdot)$ corresponding to the solution need not have a similar property since the solution need not always be stationary (even if Y_t=0,$t \in T$). Taking k=0 in (10), thus Nagabhushanam found necessary and sufficient conditions on a_j's such that the X_t-process is also stationary and the solution is unique. This is much harder in the general case. Bochner has analyzed the situation in [2] when the Y_t-process is stationary, but the solution (or input) process is weakly harmonizable. The following somewhat surprising result is a consequence of his work. A simplified proof will be given here.

Theorem 2.1 Let the output process $\{Y_t, t \in T\}$ be stationary and suppose that the input process $\{X_t, t \in T\}$ is weakly harmonizable and satisfies (7) or (8) with $C(= \tilde{C}$ or $C_1)$ as the spectral characteristic of the filter Λ. If $Q = \{\lambda : C(\lambda) = 0\}$, the zero set of C, and F_y is the spectral measure of the output process, then one has

$$\int_{\hat{T}-Q} |C(\lambda)|^{-2} F_y(d\lambda) < \infty, \tag{13}$$

and, moreover, there exists a stationary input $\{X_t^o, t \in T\}$ which satisfies $\Lambda X_t^o = Y_t^o, t \in T$, where $\{Y_t^o, t \in T\}$ is stationary whose spectrum coincides with that of $\{Y_t, t \in T\}$ on $\hat{T}-Q$. The solution X_t^o is given by

$$X_t^o = \int_{\hat{T}-Q} e^{it\lambda} C(\lambda)^{-1} Z_y(d\lambda). \tag{14}$$

Remark. The hypothesis that X_t is a solution of $\Lambda X_t = Y_t$, can be weakened, by only assuming that it is a "distributional solution". This means, for instance considering (8), that

$$\int_T \phi(t) Y_t\, dt = \int_T \Big(\sum_{j=0}^{k} (-1)^j \int_{\hat{T}} \phi^{(j)}(t-u) H_j(du)\Big) X_t\, dt, \tag{15}$$

for all $\phi(\cdot)$ having k-continuous derivatives, and ϕ has a compact support. The right side of (15) can be expressed as $\int_T X_t(\Lambda^* \phi)(t)dt$ where the formal "adjoint" Λ^* of Λ is the expression in parenthesis. It can be shown that each solution of (6) satisfies (15) so that each usual (or "strong") solution is also a distributional one. The converse holds in some special cases, but usually only under more restrictions. Thus if (7) or (8) has a distributional solution, then the constructed solution X_t^o of (14) satisfies $\Lambda X_t^o = Y_t^o, t \in T$, in the same distributional sense, i.e., satisfies the system of equations (15). The proof is given in this general form.

Proof: Suppose first that (13) is true. Then the process $\{X_t^o, t \in T\}$ by (14) is well-defined, and in fact $E(|X_t^o|^2)$ is precisely the integral in (13). That the X_t^o-process satisfies (6) with Y_t replaced by Y_t^o is seen as follows.

$$\begin{aligned} \Lambda X_t^o &= \int_{\hat{T}} C(\lambda) e^{it\lambda} Z_x^o(d\lambda), \text{ by (11) with } Z_x^o \text{ as the stochastic spectral measure of } X_t^o, \\ &= \int_{\hat{T}-Q} C(\lambda) e^{it\lambda} \frac{1}{C(\lambda)} Z_y(d\lambda), \text{ by (12) and (14)}, \\ &= \int_{\hat{T}-Q} e^{it\lambda} Z_y(d\lambda) = Y_t^o, \text{ by definition.} \end{aligned} \tag{16}$$

The spectral measures of Y_t^o and Y_t-processes clearly agree on $\hat{T}-Q$. Thus the less simple point here is a verification of (13). This is done now.

Since X_t is assumed to be a distributional solution of (6), it satisfies (15) for each ϕ on T there, when its Fourier transform ψ has two continuous derivatives and has compact support. Since both sides of (12) are elements of $L_0^2(P)$, apply an arbitrary continuous linear functional l of $L_0^2(P)$ to get

$$\int_{\hat{T}} C(\lambda) e^{it\lambda} l(Z_x(d\lambda)) = \int_{\hat{T}} e^{it\lambda} l(Z_y(d\lambda)), \text{ since } l(\cdot)$$

and the integral commute by ([10, IV.10.8(f)).

But $l(Z_x(\cdot))$ and $l(Z_y(\cdot))$ are signed measures and the above is true for all $t \in T$. Hence by the uniqueness theorem for Fourier transforms one deduces that

$$l(C(\lambda) Z_x(d\lambda)) = l(Z_y(d\lambda)).$$

Since it is true for an arbitrary continuous linear functional, this implies

$$C(\lambda) Z_x(d\lambda) = Z_y(d\lambda). \tag{17}$$

Now $C(\lambda) \neq 0$ on $\hat{T}-Q$. So one has from (17)

$$\int_{\hat{T}-Q} \frac{\psi(\lambda)}{C(\lambda)} e^{it\lambda}\, Z_y(d\lambda) = \int_{\hat{T}-Q} \psi(\lambda) e^{it\lambda}\, Z_x(d\lambda), \qquad t \in T. \tag{18}$$

But the support of ψ lies in $\hat{T}-Q$, so that in (18) $\hat{T}-Q$ may be replaced by $\hat{T}$. Then calculating the variances of these random variables, one gets

$$\int_{\hat{T}} |\frac{\psi(\lambda)}{C(\lambda)}|^2 F_y(d\lambda) = E(|\int_{\hat{T}} \psi(\lambda) e^{it\lambda} Z_x(d\lambda)|^2), \tag{19}$$

since $Z_y(\cdot)$ has orthogonal increments, $E(|Z_y(A)|^2) = F_y(A)$, and both $Z_x(A)$, $Z_y(A)$ have zero means. Using the inverse Hölder inequality on both sides of (19), if $\|\psi(\cdot)\|_\infty \leq 1$ is varied, (19) reduces to

$$\begin{aligned}\int_{\hat{T}-Q} |C(\lambda)|^{-2} F_y(d\lambda) &= \sup\{\int_{\hat{T}-Q} |\psi(\lambda)|^2 \, |C(\lambda)|^{-2} F_y(d\lambda): \|\psi\|_\infty \leq 1\} \\ &= \sup\{E(|\int_{\hat{T}} \psi(\lambda) e^{it\lambda} Z_x(d\lambda)|^2: \|\psi\|_\infty \leq 1\} \\ &\leq \sup\{(\|\psi(\cdot)e^{it(\cdot)}\|_\infty \|Z_x\|(\hat{T}))^2 : \|\psi\| \leq 1\}, \text{ by ([10], IV.10.8(c))}, \\ &\leq \|Z_x\|(\hat{T})^2 < \infty. \end{aligned} \tag{20}$$

Here $\|Z_x\|(\hat{T})$ is the semi-variation of $Z_x(\cdot)$ which is finite, (cf. [10], IV.10.2). This completes the proof.

An important consequence of this result is that if the output process $\{Y_t, t \in T\}$ is stationary, and if there is a weakly harmonizable (distributional) solution process $\{X_t, t \in T\}$ for (7) or (8), then, after deleting from the spectrum of the Y_t-process those "frequencies" which correspond to the zeros of the spectral characteristic, one can find a stationary (distributional) solution that satisfies the filter equation. Thus if the output process is not stationary but only harmonizable, then one has to analyze the filter problem further and study the input process since (19) does not hold and the construction of X_t^o by (14) fails. This general problem will be taken up in Section 6 below.

The last part of the above computation leading to (20) contains a characterization of weakly harmonizable processes implicitly, and it will be useful to separate it.

Proposition 2.2. Let $\{X_t, t \in T\}$ be a second order mean-continuous process with zero expectations. Then it is weakly harmonizable iff

$$\begin{aligned} &\text{(i)} E(|X_t|^2) \leq M < \infty, \quad t \in T, \\ &\text{(ii)} E(|\int_T \phi(t) X_t \, dt|^2) \leq C \|\hat{\phi}\|_\infty^2, \end{aligned} \tag{21}$$

for all summable functions ϕ on T (the integral is a sum if $T = \mathbb{Z}$), the $\hat{\phi}$ being the Fourier transform of ϕ. Thus X_t admits a stochastic integral representation (4) iff the above two conditions hold.

The proof that each process, representable as (4), must satisfy the above conditions (i) and (ii) is an elaboration of the calculation in (20). In fact, (i) follows immediately from the last step of (20), and for (ii) let ϕ be any such function and consider the left side of (21). Then with $T = \mathbb{R}$,

$$\begin{aligned}\int_{\mathbb{R}} \phi(t) X_t \, dt &= \int_{\mathbb{R}} \phi(t) (\int_{\mathbb{R}} e^{it\lambda} Z_x(d\lambda)) dt \\ &= \int_{\mathbb{R}} (\int_{\mathbb{R}} e^{it\lambda} \phi(t) dt) Z_x(d\lambda), \text{ the interchange being valid,}\end{aligned}$$

$$= \int_{\mathbb{R}} \hat{\phi}(\lambda) Z_x(d\lambda), \hat{\phi} \text{ being the Fourier transform of } \phi.$$

Consequently,

$$E(|\int_{\mathbb{R}} \phi(t) X_t \, dt|^2) = E(|\int_{\mathbb{R}} \hat{\phi}(\lambda) Z_x(d\lambda)|^2)$$

$$\leq \|\hat{\phi}\|_\infty^2 \|Z_x\|^2(\mathbb{R})$$

$$= C \|\hat{\phi}\|_\infty^2,$$

using the reasons as in (20). Thus (ii) holds. The converse is similar but somewhat involved. It will be omitted here.

The condition (21) was called "V-boundedness" by Bochner. He introduced these weakly harmonizable processes (but not their integral representation (4)) under the above name and showed that they are more general than the strongly harmonizable class introduced earlier by Loève [18]. A detailed discussion of these processes together with a complete proof of the above proposition may be found in [29].

In view of the preceding work, an analysis of harmonizable processes takes on added importance. Before turning to model (9), it is useful to consider the spectral domain of harmonizable processes. This will be done in the next section, which forms a background for the signal extraction problem in the following section.

III. THE SPECTRAL DOMAIN

Let $\{X_t, t \in T\}$ be a weakly harmonizable process whose covariance therefore has the representation (2) with its spectral function F having a finite Fréchet variation. Then the *spectral domain* $L^2(F)$ of the process is defined as:

$$L^2(F) = \{f : \hat{T} \to \mathbb{C} : \int_{\hat{T}} \int_{\hat{T}} f(s) \overline{f(t)} F(ds, dt) = (f, f)_F < \infty\}. \tag{22}$$

The functional $\|\cdot\|_F : f \to [(f, f)_F]^{1/2}$ is a semi-norm and $(f, g)_F$ is a semi-inner product with respect to which $L^2(F)$ becomes a semi-inner product space. If the process is stationary, then it is known for a long time that $(L^2(F), \|\cdot\|_F)$ is a complete space. However, the completeness question for the harmonizable case was left open. Even the strongly harmonizable case has not been discussed in the literature. The problem becomes much more complicated if $f : \hat{T} \to \mathbb{C}^n, n \geq 2$, i.e., if the process is multivariate. Recently a positive solution for all these processes even in the multivariate case is obtained and an aspect of it is as follows:

Theorem 3.1. Let $\{X_t, t \in T\}$ be a (scalar) weakly harmonizable process. Then its spectral domain $(L^2(F), \|\cdot\|_F)$ is a complete (semi-) inner product space. Thus it is a Hilbert space if its equivalence classes are considered.

The proof of this result, even in the multivariate case, is noted in [30] and is somewhat involved. Difficulties arise from the fact that F is not a function of one variable and also (in the weakly harmonizable case) it is not necessarily of bounded variation. It gives a "bimeasure" (cf. [20]), instead of a measure, and so an indirect route is needed. The basic ideas of the argument will be sketched.

Just as in the proof of (13), the first step here is to bypass the computations with bimeasures, and show that each weakly harmonizable process is of Karhunen class. Recall that a second order mean zero process $\{X_t, t\in T\}$ is of *Karhunen class* relative to a family $\{f_t, t\in T\}$ of Borel functions f_t: $\hat{T} \to \mathbb{C}$ and a (σ-finite) Borel measure μ on $\hat{T}$, if its covariance function $r(\cdot,\cdot)$ is expressible as:

$$r(s,t) = \int_{\hat{T}} f_s(\lambda)\overline{f_t(\lambda)}\mu(d\lambda), \qquad s,t \in T. \tag{23}$$

If $f_s(\lambda) = e^{is\lambda}$, then this becomes the stationary class. Such a process can be represented by a stochastic integral:

$$X_t = \int_{\hat{T}} f_t(\lambda)Z(d\lambda), \qquad t \in T, \tag{24}$$

where $Z(\cdot)$ is an orthogonally scattered measure on the Borel σ-algebra of $\hat{T}$. On the other hand it is a relatively deep fact that each harmonizable process can be dilated to a stationary process (cf., e.g., Niemi [23], and for a more detailed discussion [29]; see also [6], [28]). This means there exists a super Hilbert space $L_0^2(P')$, containing $L_0^2(P)$ of the given process $\{X_t, t\in T\}$, and a stationary process $\{Y_t, t\in T\}$ in $L_0^2(P')$ such that

$$X_t = QY_t, \qquad t \in T, \tag{25}$$

where Q is the orthogonal projection of $L_0^2(P')$ onto $L_0^2(P)$. But by (4), every stationary process Y_t can be expressed as

$$Y_t = \int_{\hat{T}} e^{it\lambda}\tilde{Z}(d\lambda), \qquad t \in T,$$

where $\tilde{Z}(\cdot)$ is orthogonally scattered in $L_0^2(P')$. Let $\tilde{\mu}(A) = E(|\tilde{Z}(A)|^2)$, so that $\tilde{\mu}$ is a finite Borel measure on $\hat{T}$. By a theorem of Kolmogorov (cf. [32], p. 33) there exists an orthogonal projection Π on $L^2(\tilde{\mu})$ such that

$$\begin{aligned} X_t &= QY_t = Q(\int_{\hat{T}} e^{it\lambda}\tilde{Z}(d\lambda)) \\ &= \int_{\hat{T}} \Pi(e^{it(\cdot)})(\lambda)\, \tilde{Z}(d\lambda) \qquad t \in T \end{aligned} \tag{26}$$

Let $f_t = \Pi(e^{it(\cdot)})$. Then $f_t:\hat{T} \to \mathbb{C}$ is a Borel function, $f_t \in L^2(\tilde{\mu})$ so that

$$X_t = \int_{\hat{T}} f_t(\lambda)\tilde{Z}(d\lambda), \qquad t \in T,$$

and

$$r(s,t) = E(X_s\overline{X_t}) = \int_{\hat{T}} f_s(\lambda)\overline{f_t(\lambda)}\tilde{\mu}(d\lambda), \qquad s,t \in T. \tag{27}$$

Thus the harmonizable process is indeed a Karhunen process relative to $\{f_t, t\in T\} \subset L^2(\tilde{\mu})$. The space $L^2(\tilde{\mu})$, is the spectral domain of the Y_t-process and, being the ordinary Lebesgue space of equivalence classes of square integrable functions, is complete.

The next step is to consider the spectral domain $L^2(F)$ of the weakly harmonizable space given by (22), and set up an isometric isomorphism between this space and the closed subspace $\Pi(L^2(\tilde{\mu}))$ where Π is the projection obtained in (26). The required mapping is then the correspondence

$$\iint_{\hat{T}\hat{T}} e^{is\lambda - it\lambda'} F(d\lambda,\, d\lambda') = (e^{is(\cdot)}, e^{it(\cdot)})_F$$

$$= r(s,t) = (\Pi(e^{is(\cdot)}), \Pi(e^{it(\cdot)}))_{\tilde{\mu}}.$$

Thus $e^{is(\cdot)} \rightarrow \Pi(e^{is(\cdot)})$ gives the isometry between $L^2(F)$ and the above noted closed subspace of $L^2(\tilde{\mu})$, by extending it linearly onto the space determined by the trigonometric polynomials. Here some elementary properties of the special bimeasure F must be utilized in showing that the mapping is onto. This will finally yield the completeness of $L^2(F)$ asserted in the theorem.

A consequence of the above sketch is the following inclusion relations among the various classes of time series introduced in this study:

Stationary class $\subset$ Strongly harmonizable class

$\subset$ Weakly harmonizable class

$\subset$ Karhunen class.

It should be noted at this point that, though the Karhunen class has a considerable flexibility in its definition, one has *less* information about the structural properties of its "accompanying family" $\{f_t, t\in T\}$ appearing in (23) and (24). Even the f_t of (26), although of theoretical interest, do not offer much insight in the analysis of the X_t itself. On the other hand, the harmonizable case is closely related to the Fourier analysis of bimeasures and in some ways can bring in the powerful analytic machinery of the harmonic analysis. For this reason, one turns again to the harmonizable case and uses the Karhunen representation as a technical tool in its study. For an extension and a related discussion on this class, one may refer to [7].

The above theorem as well as the remarks extend if the time series is a multiple harmonizable process. However, the questions of ranks of matrices intervene and further complications arise. It uses a special bilinear vector integration theory. Because of this, the vector case is not considered in this paper.

IV. SIGNAL EXTRACTION

Let us now consider the main signal extraction or estimation question for harmonizable processes. The model here is motivated by (9) and is stated as:

$$X_t = Y_t + Z_t, \qquad t \in T, \tag{28}$$

where $\{Y_t, t\in T\}$ and $\{Z_t, t\in T\}$ are respectively the (stochastic) signal and noise harmonizable processes which are either uncorrelated or harmonizably correlated. Here the observed process is $\{X_t, t\in T\}$ (note the change of notation from Section 2), and it is desired to estimate the signal Y_t at $t=a(\in T)$, based on one realization provided only the covariance structure of the signal and noise processes is assumed known from prior information.

Thus using the notation of (2), let F_y, F_z and F_{yz} be the spectral functions of Y_t, Z_t processes and of the harmonizable cross-covariance $r_{yz}(s,t) = E(Y_s \bar{Z}_t)$. If $F_y(\cdot,\hat{T})$, $F_z(\cdot,\hat{T})$ and $F_{yz}(\cdot,\hat{T})$ are the "marginal" spectral measures, let $h(\cdot)$, $k(\cdot)$ be defined as

$$h(\lambda) = F_y(\lambda,\hat{T}) + F_z(\lambda,\hat{T}) + F_{yz}(\lambda,\hat{T}) + \bar{F}_{zy}(\lambda,\hat{T}), \tag{29}$$

$$k(\lambda) = F_y(\lambda,\hat{T}) + F_{yz}(\lambda,\hat{T}).$$

Note that $F_{yz}(A,B) = \bar{F}_{zy}(B,A)$. The problem of obtaining the best least-squares estimator $\hat{Y}_a$ of Y_a, based on one realization of $\{X_t, t \in T\}$ whose stochastic spectral measure $Z_x(\cdot)$ is given by (4), can be presented as follows. In order to minimize the fussy detailed justifications below, all the processes in this section will be *strongly harmonizable.* With this one has:

Theorem 4.1. Let the signal plus (correlated) noise process be given by the model (28). *For any fixed* $a \in T$, *the least squares optimal signal estimator* $\hat{Y}_a$ *of* Y_a, *based on one realization of* $\{X_t, t \in T\}$, *is obtainable as:*

$$\hat{Y}_a = \int_{\hat{T}} g_a(\lambda) Z_x(d\lambda) \tag{30}$$

where g_a *is the unique solution in* $L^2(F_x)$ *of the integral equation*

$$\int_{\hat{T}} g_a(\lambda) h(d\lambda) = \int_{\hat{T}} e^{ia\lambda}\, k(d\lambda), \tag{31}$$

the $h(\cdot), k(\cdot)$, *and* $Z_x(\cdot)$ *being the set functions defined by* (29) *and* (4). *If* h, k *have densities* $h'(\cdot)$ *and* $k'(\cdot)$ *then*

$$g_a(\lambda) = k'(\lambda) e^{ia\lambda}/h'(\lambda), \qquad \text{for } almost\ all\ (\lambda). \tag{32}$$

Proof. Let $\mathbb{H}_x$ be the closed linear span of $\{X_t, t \in T\}$ in $L_0^2(P)$. The problem is to find a $\hat{Y}_a$ in $\mathbb{H}_x$ such that

$$E(|Y_a - \hat{Y}_a|^2) = \|Y_a - \hat{Y}_a\|^2 = \min\{\|Y_a - \xi\| : \xi \in \mathbb{H}_x\}.$$

Since $\mathbb{H}_x$ is closed and convex in $L_0^2(P)$, there is a unique $\hat{Y}_a$ satisfying the above condition. Moreover, the geometry of a Hilbert space implies that $\hat{Y}_a$ is simply the orthogonal projection of Y_a onto $\mathbb{H}_x, \hat{Y}_a = QY_a$. Hence $Y_a - \hat{Y}_a$ is orthogonal to $\mathbb{H}_x$, and this means, with the inner product notation,

$$(Y_a - \hat{Y}_a, X_t) = 0, \qquad t \in T. \tag{33}$$

Using (4), one notes that each element w of $\mathbb{H}_x$, being the limit - in - mean of elements of the form $\sum_{i=1}^{n} a_i^n X_{t_i}$, can be written as:

$$w = \int_{\hat{T}} v(\lambda) Z_x(d\lambda), v \in L^2(F_x).$$

Hence

$$\hat{Y}_a = \int_{\hat{T}} g_a(\lambda) Z_x(d\lambda)$$

for some $g_a \in L^2(F_x)$. Using (33) and (28), one has

$$(Y_a, X_t) = (\hat{Y}_a, X_t), \qquad t \in T,$$

so that using (2), (4) and the above expression for $\hat{Y}_a$, it follows that

$$\int_{\hat{T}}\int_{\hat{T}} e^{ia\lambda - it\lambda'} F_{yx}(d\lambda, d\lambda') = \int_{\hat{T}}\int_{\hat{T}} g_a(\lambda) e^{-it\lambda'} F_x(d\lambda, d\lambda'). \tag{34}$$

But (28) also implies $F_{yx} = F_y + F_{yz}$ and $F_x = F_y + F_z + F_{yz} + \bar{F}_{zy}$. With this one can simplify (33), after setting $t=0 (\in T)$ there, to get:

$$\int\!\!\int_{\hat{T}\hat{T}} g_a(\lambda)(F_y(d\lambda,d\lambda') + F_z(d\lambda,\, d\lambda') + 2\,\mathrm{Re}\, F_{yz}(d\lambda,d\lambda'))$$

$$= \int\!\!\int_{\hat{T}\hat{T}} e^{ia\lambda}(F_y(d\lambda,d\lambda') + F_{yz}(d\lambda,d\lambda')). \tag{35}$$

Interchanging the order of integration, which is legitimate, one has with (29):

$$\int_{\hat{T}} g_a(\lambda)h(d\lambda) = \int_{\hat{T}} e^{ia\lambda}\, k(d\lambda). \tag{36}$$

This gives the main part.

If h' and k' exist, let $g_a(\cdot)$ be defined by (32). To see that it is in $L_2(F_z)$, note that on the sets $[|g_a| > N]$ and $[|g_a| < \epsilon]$ for $0 < \epsilon < N < \infty$, with ϵ small and N large, g_a is bounded. It is close to 0 and 1 for suitable ϵ and N. On the set $[\epsilon < |g_a| < N]$ the integrand of (36) is again bounded so that g_a is in $L^2(F_z)$. Substitution of (32) in (36) shows that it is satisfied, and thus the result holds as stated.

In the case that the processes are stationary (and stationarily correlated) then F_z, F_y, F_z, F_{yz} are all supported on the diagonal $\lambda = \lambda'$ of $\hat{T}\times\hat{T}$. Thus (29) reduces to

$$h(\lambda) = F_y(\lambda,\lambda) + F_z(\lambda,\lambda) + F_{yz}(\lambda,\lambda) + \bar{F}_{zy}(\lambda,\lambda). \tag{37}$$

$$k(\lambda) = F_y(\lambda,\lambda) + F_{yz}(\lambda,\lambda).$$

Setting $\tilde{F}_y(\lambda) = F_y(\lambda,\lambda)$ etc., the integral equation (35) or (36) simplifies. Hence the solution (32), in case their densities exist, becomes

$$g_a(\lambda) = \frac{(\tilde{F}_y{}'(\lambda) + \tilde{F}_{yz}{}'(\lambda))e^{ia\lambda}}{\tilde{F}_y(\lambda) + \tilde{F}_z{}'(\lambda) + 2\mathrm{Re}\tilde{F}_{yz}{}'(\lambda)}. \tag{38}$$

In this form, the solution was first obtained by Grenander ([12], p. 275). The method of computation is a modification of the classical one (cf., e.g., Yaglom [33], p. 128). *Expected error of estimation.* In both the above cases, the expected squared error can be computed using g_a of (35) or (36). Thus

$$\begin{aligned}
\sigma_a^2 &= E(|Y_a - \hat{Y}_a|^2) \qquad (39)\\
&= E((Y_a - \hat{Y}_a)(Y_a - \hat{Y}_a)^-)\\
&= E(|Y_a|^2) + E(|\hat{Y}_a|^2) - E(\bar{Y}_a\hat{Y}_a) - E(\bar{\hat{Y}}_a Y_a)\\
&= E(\,|Y_a\,|^2) + E(\,|\,\hat{Y}_a\,|^2) - 2\,\mathrm{Re}\,E(\,|\,\hat{Y}_a\,|^2),\ \text{since } Y_a - \hat{Y}_a \ \text{ is orthogonal to } \hat{Y}_a,\\
&= E(|Y_a|^2) - E(|\hat{Y}_a|^2)\\
&= r_y(0,0) - \int\!\!\int_{\hat{T}\hat{T}} g_a(\lambda)\, \bar{g}_a(\lambda')F_z(d\lambda,d\lambda').
\end{aligned}$$

In the stationary case, if the spectral densities exist, then

$$\sigma_a^2 = \int_{\hat{T}} F_y{}'(\lambda)d\lambda - \int_{\hat{T}} |g_a(\lambda)|^2\, F_z{}'(\lambda)d\lambda.$$

Suppose $\hat{k}$ is the Fourier transform of k. Then (36) can be expressed as:

$$\int_{\hat{T}} g_a(\lambda)h'(\lambda)d\lambda = \hat{k}(a).$$

If $T = \mathbb{R}$, then this becomes the equation

$$\int_{-\infty}^{\infty} g_a(\lambda)h'(\lambda)d\lambda = \hat{k}(a). \tag{40}$$

When $\hat{k}$ and h' are known, g_a is to be solved in this equation. This is a form of the Wiener-Hopf integral equation about which much is known in the literature.

The preceding work extends to the multivariate case as well as to the weakly harmonizable series. These will not be treated here because of notational and other problems noted before.

In order to use formula (39) for a calculation of the optimal estimator $\hat{Y}_a$, one still has to find the stochastic spectral measure $Z_x(\cdot)$. However, this can be obtained with the dilation result noted for (23), and the classical results on stationary processes. This may be given as follows. Let $Z_x(\lambda)$ stand for $Z_x(\{\mu\in\hat{T}:\mu < \lambda\})$, $\lambda\in\hat{T}$. Then for $\lambda_1 < \lambda_2$ in $\hat{T}$, one has

$$\frac{Z_x(\lambda_2+) + Z_x(\lambda_2-)}{2} - \frac{Z_x(\lambda_1+) + Z_x(\lambda_1-)}{2} =$$

$$\underset{\alpha}{\text{l.i.m.}} \int_{-\alpha}^{\alpha} \frac{e^{-it\lambda_2} - e^{-it\lambda_1}}{-it} X_t \, dt, \tag{41}$$

where *l.i.m.* is the limit-in-mean (= quadratic limit), and the integral is replaced by an appropriate sum if $T=\mathbb{Z}$ (cf., e.g. [29], p. 340 and [32], p. 27). With this (39) can be used in practical problems, through efficient (algorithmic) methods are yet to be devised.

V. CLASS (KF) AND HARMONIZABILITY

Second order processes satisfying condition (3) were called class (KF) in Section 1, and they were introduced to study certain periodic behavior of some not necessarily stationary processes. Evidently the latter processes belong to class (KF). It is not immediate but true that strongly harmonizable processes are included in this class. Indeed, in the discrete case for instance, one has

$$r_n(h) = \frac{1}{n}\sum_{k=0}^{n-h-1}\int_{-\pi}^{\pi}\int_{-\pi}^{\pi} e^{ik\lambda - i(k+h)\lambda'} F(d\lambda, d\lambda'), h \geq 0, \tag{42}$$

$$= \int_{-\pi}^{\pi}\int_{-\pi}^{\pi} e^{-ih\lambda'} \frac{1-e^{i(n-h)(\lambda-\lambda')}}{n(1-e^{i(\lambda-\lambda')})\ F(d\lambda, d\lambda')},$$

$$\to \int_{-\pi}^{\pi} e^{-h\lambda}\ G(d\lambda) = r(h), \text{ as } n \to \infty,$$

where $G(\lambda) = F(\lambda,\lambda)$ and the bounded convergence theorem is used. This is possible since F is of bounded variation, and the integrals are in Lebesgue's sense. Thus $r(\cdot)$ is a stationary covariance. A similar computation (with additional arguments) shows that the same holds in the continuous parameter case also ($T=\mathbb{R}$).

If F is only of Fréchet variation finite, so that the process is weakly harmonizable, then the integral is not in Lebesgue's sense and the exchange of limit and integral in (42) is not necessarily possible. One can actually find examples for which the above limit does not exist so that the weakly harmonizable family is not included in class (KF), and a clear distinction between strong and weak classes again emerges. The following example illustrates this.

Let $\{f_n, n \in \boldsymbol{Z}\}$ be a complete orthonormal sequence in $L_0^2(P)$ where the latter space is assumed separable. Then as noted before, for each bounded linear mapping $A: L_0^2(P) \to L_0^2(P)$, the new process $Y_n = Af_n$, $n \in \boldsymbol{Z}$, is weakly harmonizable. Now choose A such that

$$Y_n = Af_n = a_n f_n = \sum_{k=0}^{\infty}(\chi_{C_k} + 2\chi_{D_k})(n) f_n, n > 0, a_0 = 1, \tag{43}$$

and let $a_{-n} = a_n$. Here C_n, D_n are the semi-closed disjoint intervals given by $C_n = [2^{2n}, 2^{2n+1}), D_n = [2^{2n+1}, 2^{2n+2})$. Thus $1 \leq a_n \leq 2$, and one finds that the Y_n-sequence is uncorrelated. But a simple computation shows that their covariance does not satisfy the limit conditions (3), since the arithmetical mean limit for variances does not exist. Thus class (KF) and weakly harmonizable families do not include each other. Such an example is not yet available in the continuous parameter case. From (42) and (43) it follows that the dominated convergence theorem, in general, fails for the integral in (2) when F is only of finite Fréchet variation.

Since the covariance $r(\cdot)$ of (3), when the limits exist, is stationary the class (KF) was termed "asymptotically stationary" by Parzen [24], who considered some properties and applications. Using this relationship, and employing the spectral representation of such $r(\cdot)$, it is possible to analyze some structure of processes in this class. Certain tentative proposals on extending the stationary prediction theory to this class have been put forward by Rozanov [31]. However, a serious study of these processes has not yet been undertaken, and this class appears to have promise. The original work of Kampé de Feriet and Frenkiel [14] on a periodic (but not stationary) class is also interesting.

VI. PHYSICAL REALIZABILITY OF FILTERS

The work of Section 2, and especially Theorem 2.1, raises two questions. If $\Lambda X_t = Y_t, t \in T$, and the output process $\{Y_t, t \in T\}$ is harmonizable, then under what restrictions on Λ (or on its spectral characteristic) does the input process $\{X_t, t \in T\}$ belong to the harmonizable class? A solution (or input-) process is *physically realizable* if it depends only on the past and present values of the output. Thus the next question is this: If the solution exists under what additional conditions is it physically realizable? The answers to these questions are obviously important for applications, and these are treated here if Λ is a difference operator.

Thus let Λ be of the kth order so that $T = \boldsymbol{Z}$ and

$$Y_n = \Lambda X_n = \sum_{l=1}^{k} a_l X_{n-l}, \qquad a_l \in \mathbb{C} \tag{44}$$

Here the constants (or weights) a_l do not depend on the time parameter "n". Assuming that $\{Y_n, n \in \boldsymbol{Z}\}$ is harmonizable the spectral characteristic of Λ in this case, for a harmonizable input X_n, is obtained from:

$$\int_{-\pi}^{\pi} e^{in\lambda} Z_y(d\lambda) = \int_{-\pi}^{\pi} \tilde{C}(\lambda) e^{in\lambda} Z_x(d\lambda), \text{ by (4)}, \tag{45}$$

where $\tilde{C}(\lambda) = \sum_{j=1}^{k} a_j e^{-ij\lambda}$. The zeros of $\tilde{C}(\cdot)$ play an important role in this analysis. Conditions for the existence and physical realizability of a solution to (44) are given in the following:

Proposition 6.1. Let the output $\{Y_n, n \in Z\}$ *be strongly harmonizable and satisfy* (44). *Let* $Q = \{\lambda : \tilde{C}(\lambda) = 0\} \subset \hat{T} = (-\pi, \pi]$. *Then a strongly harmonizable solution of* (44) *exists iff* (i) $|F_y|(Q \times Q) = 0$, *and* (ii)

$$\int_{Q^c}\int_{Q^c} |\tilde{C}(\lambda)\bar{\tilde{C}}(\lambda')|^{-1} |F_y|(d\lambda, d\lambda') < \infty, \tag{46}$$

where $|F_y|$ *is the variation measure of the spectral function* F_y *of the output series (which exists). Moreover, if the roots of the characteristic equation*

$$p_k(z) = \sum_{j=0}^{k} a_j z^j = 0$$

lie outside the unit circle of the complex plane, then the filter is physically realizable so that

$$X_n = \sum_{j=0}^{\infty} b_j Y_{n-j} \tag{47}$$

where the b_j *are coefficients of* $(p_k(z))^{-1}$ *when expanded in a power series around each root of the equation* $p_k(z) = 0$.

Proof. The essential argument will be sketched here since it exhibits the structure of the problem. Equation (18) reduces in the present case to:

$$X_n = \int_{Q^c} e^{in\lambda}(\tilde{C}(\lambda))^{-1} Z_y(d\lambda), \qquad n \in \mathbf{Z}, \tag{48}$$

where (i) is used. That X_n satisfies (44) follows from Theorem 2.1 itself, since (46) is essentially the same as (13) there.

Regarding the converse, by (45) one gets easily after squaring and taking expectations, the following:

$$\int_A\int_B |F_y|(d\lambda, d\lambda') = \int_A\int_B \tilde{C}(\lambda)\bar{\tilde{C}}(\lambda')|F_x|(d\lambda, d\lambda') \tag{49}$$

for all Borel sets A, B of $\hat{T}$. In particular, if $A = B = Q$ and noting that $\tilde{C}$ vanishes on Q, one gets (i). Next taking $A = B = Q^c$ in (49) and noting that $\tilde{C}(\lambda) \neq 0$ on Q^c one has

$$\int_{Q^c}\int_{Q^c} (|\tilde{C}(\lambda)\bar{\tilde{C}}(\lambda')|)^{-1} |F_y|(d\lambda, d\lambda') = \int_{Q^c}\int_{Q^c} |F_x|(d\lambda, d\lambda') < \infty.$$

So (46) holds.

When (i) and (ii) are satisfied, then X_n is given by (48). Since $\tilde{C}(\lambda) = p_k(e^{-i\lambda})$, and the zeros of p_k are now assumed to lie outside the unit circle, one can expand $(p_k(z))^{-1}$ in power series on using the partial fractions. Collecting the like terms, one gets

$$(p_k(z))^{-1} = \sum_{j=0}^{\infty} b_j z^i,$$

which is uniformly and absolutely convergent. Hence (48) becomes

$$X_n = \int_{Q^c} e^{in\lambda} \sum_{j=0}^{\infty} b_n e^{ij\lambda} Z_y(d\lambda)$$

$$= \sum_{n=0}^{\infty} b_j Y_{n-j},$$

so that X_n depends only on the past and present of the output $\{Y_n, n \in \mathbf{Z}\}$.

This result in the stationary case was obtained in [21] and the latter work was extended by Kelsh [16] to the strongly harmonizable case. Further conditions must be imposed for the uniqueness of the solution. If $Q=\phi$, then there is uniqueness. In general, however, among all harmonizable solutions of (44), there is only one whose spectral measure F_x satisfies $|F_x|(Q\times Q)=0$. This is not a satisfactory situation since when $Q\neq\phi$, it is not easy to test all the solutions for this condition. But no other usable condition has yet been found. Simple examples were exhibited in [16] to show how the situation changes on this point from the stationary series treated in [21].

The weakly harmonizable case also admits a similar type characterization, but conditions (i) and (ii) must be replaced by those on its stochastic spectral measure since $|F_x|, |F_y|$ need not be finite. The corresponding statement reads as follows.

Proposition 6.2 Let $\{Y_n, n\in\mathbf{Z}\}$ be weakly harmonizable and consider (44). *Then there exists a weakly harmonizable solution process iff* (i) $\|Z_y\|(Q)=0$, *where* $\|Z_y\|(\cdot)$ *is the semi-variation of the stochastic spectral measure* $Z_y(\cdot)$ *of the output series* Y_n, *and* (ii) $\int_{Q^c}(\tilde{C}(\lambda))^{-1}Z_y(d\lambda)$ *exists, as a Dunford-Schwartz integral. Further, the filter is physically realizable whenever the roots of the characteristic equation* $p_k(z)=0$ *lie outside the unit circle, as before.*

This result and its multivariate extension were given by Chang [5]. The details and other related work will be given in a later publication. Results if Λ is a differential and an integro-differential operator are also possible, and some aspects of these have been studied. However, the best or minimal conditions and computational methods are still not available.

The preceding work depends essentially on the fact that the filter coefficients in (7), (8) or (44) are independent of time. If they are not constant, two things can happen: (i) they depend on time but are still nonrandom, or (ii) they are random and perhaps also involve time. In the latter case, specially if the time parameter is continuous ($T=\mathbb{R}$) then the work leads to stochastic differential equations and results in *nonlinear* filtering problems. In the former case, the filters are linear but, being time dependent, the present point of view must be replaced by a different approach. An indication of the kind of resulting work may be seen from the early studies of Dolph and Woodbury [9], and it is potentially useful here. Certain other related statistical questions on time series, especially when $T=\mathbb{R}$, are outlined in the interesting survey of Cramér [8]. Both of these are useful problems which deserve a careful study in the future. These processes are almost always nonstationary and seldom Gaussian. But they have enough structure to be treated with the presently available tools.

VII. SAMPLING THE PROCESS

In the study of harmonizable (or for that matter any other) processes, a problem of interest for applications is to sample the process and then deduce its properties from the sample. There are two standard methods for this purpose. One is to sample at a bounded set of points, the other is to sample at periodic times. Since, especially in the second case, there can be two different processes which may agree at these fixed equidistant points (called the "aliasing" problem), it is necessary to choose the distance between the observations small enough so that the above "aliasing" is avoided. Thus different sets of conditions are needed for these two types of sampling. The following results show how these cases may be successfully handled.

Proposition 7.1 Let $\{X_t, t \in \mathbb{R}\}$ be a strongly harmonizable process with its spectral function F_x for which the moment generating function exists, in the sense that

$$\int_{\mathbb{R}} \int_{\mathbb{R}} e^{s\lambda + t\lambda'} |F_x|(d\lambda, d\lambda') < \infty, \qquad s, t \in \mathbb{R}, \tag{50}$$

where $|F_x|$ is the variation measure of F_x. Then for any bounded infinite set of distinct points $\{t_i, i \geq 1\} \subset \mathbb{R}$, the sample $\{X_{t_i}, i \geq 1\}$ determines the given process on the whole line.

Proof. Let $r(\cdot,\cdot)$ be the covariance function of the process. Then (50) implies that r is continuously differentiable in $\mathbb{R} \times \mathbb{R}$, because of the representation (2) for (strongly) harmonizable processes. Since the set $\{t_i, i \geq 1\}$ is bounded and infinite, it has a convergent subsequence with limit t_0 (say) in $\mathbb{R}$. These two facts imply from ([19], p. 471) that X_t is an analytic random function in mean-square and can be expanded in a neighborhood of t_0 as:

$$X_t = \sum_{n=0}^{\infty} \frac{(t-t_0)^n}{n!} X_{t_0}^{(n)}, \tag{51}$$

where $X_{t_0}^{(n)}$ is the nth derivative of X_t at t_0 in the mean-square sense (as in (7)). Since t_0 is a limit point of an infinite subset of the given $\{t_i, i \geq 1\}$, denoted by the same index for ease, it follows that the process and $\{X_{t_i}, i \geq 1\}$ agree in this neighborhood, and by the analyticity of the process, the function defined by (51) and the given one also agree in the whole neighborhood. However, the domain of these functions is $\mathbb{R}$, a connected set, and the process regarded as a function from $\mathbb{R}$ into $L_0^2(P)$ extends the one given by (51), which is also analytic. But then by the classical principle of analytic continuation (cf., [3], p. 40, and see [13] p. 98 together with Theorem 3.14.1), the process $\{X_t, t \in \mathbb{R}\}$ is the only such which agrees with $\{X_{t_i}, i \geq 1\}$. This is precisely the assertion of the proposition.

Relaxing the moment condition (50), but using the carrier of F_x, one can present a sufficient condition for the periodic sampling case. Recall that the *carrier* or *support* of F_x (in $\mathbb{R}^2$) is the smallest closed set S such that for every open set $A \subset S^c$ one has $|F_x|(A) = 0$ or more generally $\sup_{A \subset S^c} |F_x(A)| = 0$ where the supremum is taken over all open sets in S^c. Both are equivalent for the strongly harmonizable case, and the second one is used for the weakly harmonizable process. The following two propositions give sufficient conditions for the periodic sampling in the strong and weak harmonizable cases respectively.

Proposition 7.2 Let $\{X_t, t \in \mathbb{R}\}$ be a strongly harmonizable process with S as the support of its spectral function F_x. Then the sample $\{X_{nh}, n \in Z\}$ determines the process uniquely if the carrier is a "wandering set" under all translations by h units in the following sense:

$$T_h^i(S) \cap T_h^j(S) = \phi, i \neq j, \tag{52}$$

where $T_h(S) = \{(x+h, y+h) : (x,y) \in S\}$, $T_h^{-1} = T_{-h}(S)$, and the higher powers are defined iteratively $T_h^i(S) = T_h(T_h^{i-1}(S))$, $h > 0$.

This is an extension to the harmonizable case of a result due to Lloyd [17] in the stationary case, and the proof is given in [27] which also contains a generalization of the preceding result for certain Cramér classes introduced in [7]. For the stationary case condition (52) becomes necessary, but this is not so for the harmonizable processes, (contrary to a casual remark in [27] without proof). Looking at the behavior of the carrier of F_x another, perhaps more usable, condition can be presented for the periodic sampling of even weakly harmonizable processes.

The following contains the desired extension: ($X_t = X(t)$ again).

Proposition 7.3 Let $\{X(t), t \in \mathbb{R}\}$ *be a weakly harmonizable process with its spectral function* F_x *given by* (2). *Suppose that for each* $\epsilon > 0$, *there exists a bounded set* $A_\epsilon \subset \mathbb{R}$ *such that for all Borel sets* $B \subset A_\epsilon^c$

$$\sup_{B \subset A_\epsilon^c} F_x(B,B) < \epsilon. \tag{53}$$

If σ_0 *is the diameter of the set* A_ϵ (i.e., $\sigma_0 = \sup\{|x-y| : x,y \in A_\epsilon\}$) *then for any* $h > \sigma_0$,

$$E(|X(t) - X_n(t)|^2) \leq [C(t)/n(h-\sigma_0)] + 4\epsilon, \tag{54}$$

for some constant $0 < C(t) < \infty$, *which is bounded on bounded* t- *sets and where*

$$X_n(t) = \sum_{k=-n}^{n} X(\frac{k\pi}{h}) \frac{\sin(th-k\pi)}{(th-k\pi)}. \tag{55}$$

In case the carrier S *of* F_x *is bounded, if one lets* $A_\epsilon \times A_\epsilon \supset S$, *then* (53) *is automatic and one can set* $\epsilon = 0$ *in* (54).

This result explicitly gives an error estimate in taking the specific $h > 0$ and $n \geq 1$. The estimate depends on an approximation of $e^{it\lambda}$ by suitable "sine" series in (4). The method is an extension of the one for strongly harmonizable processes given by Piranashvili [25], and the details of proof as well as a further generalization for a Cramér class can be found in [4]. Since the argument is somewhat involved, it will not be given here. This type of result for the nonstochastic, and then for stationary, processes is classical and is known as the Kotel'nikov-Shanon type sampling (cf., e.g., [11], p. 204). A multivariate stationary extension of the latter has recently been considered by Pourahmadi [26] and its generalization for harmonizable processes is possible, although it is not automatic, and the details have not been completed. Thus these are some of the problems arising from this work for future investigations.

REFERENCES

1. Bochner, S. (1953), "Fourier transforms of time series," *Proc. Nat. Acad. Sci.*, Vol. 39, 302-307.

2. Bochner, S. (1956), "Stationarity, boundedness, almost periodicity of random valued functions," *Proc. Third Berkeley Symp. Math. Statist. and Prob.*, Vol. 2, 7-27.

3. Cartan, H. (1963), *Elementary Theory of Analytic Functions of One and Several Complex Variables,* Hermann, Paris.

4. Chang, D.K., and Rao, M.M. (1983), "Bimeasures and sampling theorems for weakly harmonizable processes," *Stoch. Anal. Appl.* Vol. 1, 21-55.

5. Chang, D.K. (1983), *Bimeasures, Harmonizable Processes and Filtering,* Dissertation, University of California, Riverside, CA.

6. Chatterji, S. D. (1982), "Orthogonally scattered dilation of Hilbert space valued set functions," *(Oberwolfach Proceedings), Lect. Notes Math.* Vol. 945, 269-281.

7. Cramér, H. (1951), "A contribution to the theory of stochastic processes," *Proc. Second Berkeley Symp. Math. Statist. and Prob.*, 329-339.

8. Cramér, H. (1971), *Structural and Statistical Problems for a Class of Stochastic Processes,* Princeton University Press.

9. Dolph, C.L., and Woodbury, M.A. (1952), "On the relation between Green's functions and covariances of certain stochastic processes and its application to unbiased linear prediction," *Trans. Amer. Math. Soc.*, Vol. 72, 519-550.

10. Dunford, N., and Schwartz, J.T. (1958), *Linear Operators: Part I, General Theory,* Wiley-Interscience, New York.

11. Gikhman, I.I., and Skorokhod, A.V. (1969), *Introduction to the Theory of Random Processes,* W.B. Saunders Company, Philadelphia, PA.

12. Grenander, U. (1950), "Stochastic processes and statistical inference," *Ark. f. Mat.*, Vol. 1, 195-277.

13. Hille, E., and Phillips, R.S. (1957), *Functional Analysis and Semigroups,* Amer. Math. Soc. Collq. Publ., New York.

14. Kampé de Feriet, J., and Frenkiel, F.N. (1962), "Correlation and spectra for nonstationary random functions," *Math. Comp.*, Vol. 16, 1-21.

15. Karhunen, K. (1947), "Lineare Transformationen stationärer stochastischer Prozesse," *Den. 10 Skandinaviske Matematiker Kongres i Kobenhaven,* 320-324.

16. Kelsh, J.P. (1978), *Linear Analysis of Harmonizable Time Series,* Dissertation, University of California, Riverside, CA.

17. Lloyd, S.P. (1959), "A sampling theorem for stationary (wide sense) stochastic processes," *Trans. Amer. Math. Soc.*, Vol. 92, 1-12.

18. Loève, M. (1948), "Fonctions aléatories du second order," Note in P. Levy's *Processes Stochastique et Movement Brownien,* Gauthier-Villars, Paris, 228-352.

19. Loève, M. (1955), *Probability Theory,* D. Van Nostrand Company, Princeton, NJ.

20. Morse, M., and Transue, W. (1956), "ϕ-bimeasures and their integral extensions," *Ann. Math.*, Vol. 64, 480-504.

21. Nagabhushanam, K. (1951), "The primary process of a smoothing relation," *Ark. f. Math.*, Vol. 1, 421-488.

22. Niemi, H. (1975), "Stochastic processes as Fourier transforms of stochastic measures," *Ann. Acad. Scient. Fennicae, Ser. A I, No.* Vol. 591, 1-47.

23. Niemi, H. (1977), "On orthogonally scatterred dilations of bounded vector measures," *Ann. Acad. Scient. Fennicae, Ser. A* I. *Math.,* Vol. 3, 43-52.

24. Parzen, E. (1962), "Spectral analysis of asymptotically stationary time series," *Bull. Inst. Internat. Statist.,* Vol. 39, 87-103.

25. Piranashvili, Z. (1967), "On the problem of interpolation of random processes," *Theor. Probl. Appl.,* Vol. 7, 647-659.

26. Pourahmadi, M. (1983), "A sampling theorem for multivariate stationary processes," *J. Multivar. Anal.,* Vol. 13, 177-186.

27. Rao, M.M. (1967), "Inference in stochastic processes-III," *Z. Wahrs.* Vol. 8, 49-72.

28. Rao, M.M. (1982), "Domination problem for vector measures and applications to nonstationary processes," *(Oberwolfach Proceedings) Lect. Notes Math.,* Vol. 945, 296-313.

29. Rao, M.M. (1982), "Harmonizable processes: structure theory," *L'Enseign. Math.,* Vol. 28, 295-351.

30. Rao, M.M. (1984), "The spectral domain of multivariate harmonizable processes," *Proc. Nat. Acad. Sci.,* Vol. 82, 4611-4612.

31. Rozanov, Yu. A. (1959), "Spectral analysis of abstract functions," *Theor. Prob. Appl.,* Vol. 4, 271-287.

32. Rozanov, Yu. A. (1967), *Stationary Random Processes,* Holden-Day, Inc., San Francisco, CA.

33. Yaglom, A.M. (1962), *Introduction to the Theory of Stationary Random Functions,* Prentice-Hall, Englewood Cliffs, NJ.

FISHER CONSISTENCY OF AM-ESTIMATES OF THE AUTOREGRESSION PARAMETER USING HARD REJECTION FILTER CLEANERS

R. D. Martin * †
V. J. Yohai ** † ‡

ABSTRACT

An AM estimate $\hat{\phi}$ of the AR(1) parameter ϕ is a solution of the M-estimate equation $\sum_1^n \hat{x}_{t-1}\psi([y_t - \hat{\phi}\hat{x}_{t-1}]/s_t) = 0$ where $\hat{x}_{t-1}$, $t = 0, 2, \ldots,$ satisfies the robust filter recursion $\hat{x}_t = \hat{\phi}\hat{x}_{t-1} + s_t\psi^*([y_t - \hat{\phi}\hat{x}_{t-1}]/s_t)$, and s_t is a data dependent scale which satisfies an auxiliary recursion. The AM-estimate may be viewed as a special kind of bounded-influence regression which provides robustness toward contamination models of the type $y_t = (1 - z_t)x_t + z_t w_t$ where z_t is a 0–1 process, w_t is a contamination process and x_t is an AR(1) process with parameter ϕ. While AM-estimates have considerable heuristic appeal, and cope with time series outliers quite well, they are not in general Fisher consistent. In this paper, we show that under mild conditions, $\hat{\phi}$ is Fisher consistent when ψ^* is of hard-rejection type.

I. INTRODUCTION

In recent years several classes of robust estimates of ARMA model parameters have been proposed. The three major classes of such estimates are: (i) GM-estimates (Denby and Martin, 1979; Martin, 1980; Bustos, 1982, Künsch, 1984), (ii) AM-estimates (Martin, 1980; Martin, Samarov and Vandaele, 1983), and (iii) RA-estimates (Bustos, Fraiman and Yohai, 1984; Bustos and Yohai, 1986). See Martin and Yohai (1985) for an overview.

Each of the three types of estimates appear to have advantages over the others in certain circumstances. However, in some overall sense the AM-estimates seem most appealing: They are based in on an intuitively appealing robust filter-cleaner which "cleans" the data by replacing outliers with interpolates based on previous cleaned data. Furthermore, they have proved quite useful in a variety of applications (in addition to the references given after (ii) above, see also Kleiner, Martin and Thomson, 1979, and Martin and Thomson, 1982). On the other hand, the AM-estimates are sufficiently complicated functions of the data that it has proven difficult to establish even the most basic asymptotic properties such as consistency. Indeed, it appears that in general AM-estimates are not consistent (see the complaint of Anderson, 1983, in his discussion of Martin, Samarov and Vandaele, 1983), even though their asymptotic bias appears to be quite small (see the approximate bias calculation in Martin and Thomson, 1982).

* Professor of Statistics, Department of Statistics, GN-22, University of Washington, Seattle, Washington USA 98195

** Professor of Statistics at the Departmento de Matematica, Facultad de C. Exactas Y Naturales, Ciudad Universitaria, Pabellon 1, 1428 Buenos Aires, Argentina, and Senior Researcher at CEMA, Virrey del Pino 3210, 1428 Buenos Aires, Argentina.

† Research supported by ONR Contract N00014-84-C-0169

‡ Research supported by a J.S. Guggenheim Memorial Foundation fellowship.

In this paper we consider only a special case of AM-estimates based on a so-called *hard-rejection* filter cleaner. The importance of hard-rejection filter-cleaners, which are described in Section 2 for the first-order autoregressive (AR(1)) model, is that engineers often use a similar intuitively appealing modification of the Kalman filter for dealing with outliers in tracking problems. In Section 3 we prove that (under certain assumptions) these special AM-estimates are Fisher consistent for the parameter ϕ_0 of an AR(1) model, Fisher consistency being the first property one usually establishes along the way to proving consistency. In addition we prove uniqueness of the root of the asymptotic estimating equation, in the special circumstance where we know the sign of the true autoregression parameter.

The AR(1) model we consider is

$$x_t = \phi_0 x_{t-1} + u_t , \qquad t = 0, \pm 1, \pm 2, \cdots \tag{1.1}$$

along with the assumption

(A1) *The u_t's are independent and identically distributed with symmetric distribution F which assigns positive probability to every interval.*

Furthermore, we shall let σ denote a measure of scale for the u_t's. For example, σ might be the median absolute deviation (MAD) of the u_t, scaled to yield the usual standard deviation when the u_t are Gaussian, namely, $\sigma = MAD\big/.6795$.

A model-oriented justification for using a robust procedure such as the AM-estimates treated here is that the observations are presumed to be given by the general contamination model

$$y_t = (1 - z_t^{\gamma})\, x_t + z_t^{\gamma}\, w_t \tag{1.2}$$

where z_t^{γ} is a 0–1 process with $P(z_t^{\gamma} = 1) = \gamma + o(\gamma)$, and w_t is an outlier generating process. The processes z_t^{γ}, w_t and x_t are presumed jointly stationary. See, for example, Martin and Yohai (1986).

The filter-cleaners and AM-estimates introduced in the next section are designed to cope well with outliers generated by such a model. However, in this paper our main focus will be on the behavior of the AM-estimates only at the nominal model (1.1), i.e., when $z_t^{\gamma} \equiv 0$ in (1.2).

II. AM-ESTIMATES AND HARD-REJECTION FILTER CLEANERS

2.1 Filter Cleaners and AM-Estimates for the AR(1) Parameter

Suppose that the model (1.2) holds, for the moment with or without the condition $z_t^{\gamma} \equiv 0$.

Let $\hat{x}_t = \hat{x}_t(\phi)$ denote the filter-cleaner values generated for $t = 1, 2, \cdots$ by the robust filter cleaner recursion

$$\begin{aligned}
\hat{x}_t &= \phi\, \hat{x}_{t-1} + s_t \psi^* \left(\frac{y_t - \phi\, \hat{x}_{t-1}}{s_t} \right) \\
s_t^2 &= \phi^2 p_{t-1} + \sigma^2 \\
p_t &= s_t^2 \left[1 - w^* \left(\frac{y_t - \phi\, \hat{x}_{t-1}}{s_t} \right) \right]
\end{aligned} \tag{2.1}$$

with initial conditions

$$\hat{x}_0 = 0$$

$$s_0^2 = \frac{\sigma^2}{1-\phi^2}. \tag{2.1'}$$

The robustifying psi-function ψ^* is odd and bounded, and the weight function w^* is defined by

$$w^*(r) = \frac{\psi^*(r)}{r}. \tag{2.2}$$

We shall often use the notation, $\hat{x}_t(\phi)$ and $s_t(\phi)$ to emphasize the dependence of $\hat{x}_t$ and s_t on ϕ. Then an AM-estimate $\hat{\phi}$ of ϕ is defined by

$$\sum_{t=2}^{n} \hat{x}_{t-1}(\hat{\phi})\, \psi\left(\frac{y_t - \hat{\phi}\hat{x}_{t-1}(\hat{\phi})}{s_t(\hat{\phi})}\right) = 0 \tag{2.3}$$

where the robustifying function ψ is odd and bounded, but in general different than ψ^*. Since bounded ψ^* gives rise to bounded $\hat{x}_t$'s (see Martin and Su, 1986), the AM-estimate $\hat{\phi}$ can be regarded as a form of bounded influence regression (see Hampel et al., 1986). Let $\hat{\phi}_M$ be the "ordinary" M-estimate defined by

$$\sum_{t=2}^{n} y_{t-1} \psi\left(\frac{y_t - \hat{\phi}_M y_{t-1}}{\hat{s}}\right) = 0 \tag{2.4}$$

where $\hat{s}$ is some robust estimate of scale of the residuals $y_t - \hat{\phi}_M y_{t-1}$. The estimate $\hat{\phi}_M$ does not have bounded influence (see Martin and Yohai, 1986). The bounded influence estimate $\hat{\phi}$ defined by (2.3) is obtained from (2.4) by replacing y_{t-1} by $\hat{x}_{t-1}(\hat{\phi})$, and by replacing the global scale estimate $\hat{s}$ by the local, data-dependent scale s_t. Although the M-estimate $\hat{\phi}_M$ has high efficiency robustness at perfectly observed autoregressions (Martin, 1979), $\hat{\phi}_M$ is known to lack qualitative robustness (see for example Martin and Yohai, 1985), and the $\hat{\phi}$ of (2.3) represents a natural kind of robustification of $\hat{\phi}_M$.

We can characterize the asymptotic value of $\hat{\phi}$ as follows. First, assume that the filter recursions (2.1) are started not at $t = 0$, but in the remote past, and that $\hat{x}_t$, s_t and y_t are jointly asymptotically stationary. Then consider the equation

$$E\, \hat{x}_{t-1}(\phi(\mu))\psi\left(\frac{y_t - \phi(\mu)\hat{x}_{t-1}(\phi(\mu))}{s_t(\phi(\mu))}\right) = 0 \tag{2.5}$$

where μ is the measure for the process y_t, and the choice of t is arbitrary by virtue of starting the filter in the remote part. It is presumed that the functional $\phi(\mu)$ is well-defined by (2.5). Under reasonable conditions one expects that $\hat{\phi}$ is strongly or weakly consistent, i.e., that will converge to $\phi(\mu)$ almost surely, or in probability.

2.2 Fisher Consistency

A minimal requirement for any estimate, including robust estimates, is that of Fisher consistency. In the present context this means: when $z_t^\gamma \equiv 0$ in the general contamination model (1.2), we have $y_t \equiv x_t$ and then x_t has measure μ_{ϕ_0} where ϕ_0 is the true parameter value. Then $\hat{\phi}$ is said to be *Fisher consistent* if

$$\phi(\mu_{\phi_0}) = \phi_0 \qquad \forall\ \phi_0 \in (-1, 1). \tag{2.6}$$

In general, AM-estimates are not Fisher consistent. The plausibility of the claim is easy to see in the case where $\psi^* = \psi$. Substituting the basic filter equation of (2.1) in (2.3) gives:

$$\sum_{t=2}^{n} \hat{x}_{t-1}(\hat{\phi}) \frac{[\hat{x}_t(\hat{\phi}) - \hat{\phi}\hat{x}_{t-1}(\hat{\phi})]}{s_t} = 0 . \tag{2.7}$$

Thus, in this special case, $\hat{\phi}$ can be characterized as a weighted least squares estimate based on the cleaned data $\hat{x}_t = \hat{x}_t(\hat{\phi})$. When $y_t \equiv x_t$ is an outlier free *Gaussian* process, a properly tuned filter-cleaner will result in $\hat{x}_t = x_t$ for most, *but not all,* times t. At those times t for which $\hat{x}_t \neq x_t$, $\hat{x}_t$ will typically be more highly correlated with $x_{t-1}, x_{t-2}, \cdots,$ than is x_t. Thus, neither weighted nor classical least squares applied to the $\hat{x}_t$ is expected to yield consistent, or even Fisher consistent, estimates. This will be the case a fortiori when $y_t \equiv x_t$, but x_t has innovations outliers by virtue of the distribution of u_t having a heavy-tailed distribution (in which case the event $\hat{x}_t \neq x_t$ will occur more frequently).

The surprising result is that use of a hard-rejection filter cleaner does yield Fisher consistency under reasonable assumptions. In particular, according to our working assumption A1, the x_t process need not be Gaussian.

2.3 Hard-Rejection Filter Cleaners

From now on we take $z_t^{\gamma} \equiv 0$, and take ψ^* to be of the hard rejection type

$$\psi^*(r) = \begin{cases} r & |r| \leq c \\ 0 & |r| > c . \end{cases} \tag{2.8}$$

Correspondingly

$$w^*(r) = \begin{cases} 1 & |r| \leq c \\ 0 & |r| > c . \end{cases} \tag{2.10}$$

The constant c is adjusted to achieve a proper tradeoff between efficiency and robustness of the filter-cleaner (see Martin and Su, 1986, for guidelines here). The results in the remainder of the paper hold for any $c > 0$, and without lost of generality we take $c = 1$.

Note that when ψ^* in (2.1) is the hard-rejection type, the filter-cleaner value at time t is either $\hat{x}_t = y_t$ or $\hat{x}_t = \phi\,\hat{x}_{t-1}(\phi)$.

We can now characterize the hard-rejection filter as follows. Let the filter parameter be ϕ, and *from now on* replace y_t by x_t in (2.1). Then since $\psi^*(r)$ is either 0 or r in accordance with whether or not $|x_t - \phi\,\hat{x}_{t-1}(\phi)| \gtrless s_t$, it is easy to see that $\hat{x}_t(\phi)$ must have the form

$$\hat{x}_t(\phi) = \phi^{L_t}\, x_{t-L_t} \tag{2.11}$$

where $L_t = L_t(\phi)$ is the random time which has elapsed since the *last* "good" x_m. A "good" x_m is one for which $|x_m - \phi\,\hat{x}_{m-1}(\phi)| < s_m$, and hence $\hat{x}_m(\phi) = x_m$.

Let

$$N_t(\phi) = \text{the latest time, less or equal to } t \text{, at which a good } x_t \text{ occurs.} \tag{2.12}$$

Then

$$L_t(\phi) = t - N_t(\phi). \tag{2.13}$$

Note from (2.1) with $y_t = x_t$, that for a good x_t we have $p_t = 0$ and $s_{t+1}^2 = \sigma^2$. Let

$$K_j^* = (\sigma^2 \sum_{k=0}^{j} \phi^{2k})^{\frac{1}{2}}, \quad j = 0, 1, 2, \cdots . \tag{2.14}$$

Then $s_t^2 = (K_l^*)^2$ if and only if $L_{t-1}(\phi) = l$

Now set

$$u_t(\phi) = x_t - \phi\, \hat{x}_{t-1}(\phi) \tag{2.15}$$

and note that the event M_t^* that x_t is bad (i.e., x_t is not good) occurs if and only if $u_t(\phi)$ is "rejected", i.e., if $|u_t(\phi)|$ is larger than the appropriate K_j^*. The appropriate K_j^* is $K_{L_{t-1}(\phi)}^*$, and so we can write

$$M_t^* = [\,|u_t(\phi)| \geq K_{L_{t-1}(\phi)}^*\,]. \tag{2.16}$$

Note that

$$M_t^* = [\hat{x}_t(\phi) = \phi\, \hat{x}_{t-1}(\phi), \quad N_t(\phi) = N_{t-1}(\phi)]$$

and

$$(M_t^*)^c = [\hat{x}_t(\phi) = x_t, \quad N_t(\phi) = t\,].$$

For any j we can use (1.1) to write

$$x_t = \phi_0^j\, x_{t-j} + \sum_{k=0}^{j-1} \phi_0^k\, u_{t-k} . \tag{2.17}$$

If we set $j = L_{t-1}$ and $\phi(\mu_{\phi_0}) = \phi_0$, then (2.11) and (2.17) give

$$\begin{aligned} x_t - \phi_0\, \hat{x}_{t-1}(\phi_0) &= x_t - \phi_0^{1+L_{t-1}}\, x_{t-1-L_{t-1}} \\ &= \sum_{k=0}^{L_{t-1}} \phi_0^k\, u_{t-k} . \end{aligned}$$

In this case, with $y_t = x_t$ and $(\phi(\mu)) = \phi_0$, the left-hand side of (2.5) becomes

$$E\, \phi_0^{L_{t-1}}\, x_{t-1-L_{t-1}}\, \psi\left(\frac{\sum_{k=0}^{L_{t-1}} \phi_0^k\, u_{t-k}}{K_{L_{t-1}(\phi_0)}^*} \right). \tag{2.18}$$

Now if L_{t-1} were replaced by a fixed value m, then the independence of $u_{t-m}, \cdots, u_t$ and x_{t-m-1}, along with the evenness assumption on the distribution of the u_t and oddness assumption for ψ, would result in the above expectation being zero. This would give part of what is required to establish Fisher consistency — the other part is to show that (2.18) is non-zero when ϕ_0 is replaced by $\phi \neq \phi_0$. However, even for this first part a more detailed argument is required because $x_{t-L_{t-1}}$ and $u_{t-L_{t-1}}, \cdots, u_t$ are *not* conditionally independent, given $L_{t-1} = m$. Fortunately, symmetry and skewness arguments presented in the next section allow one to get around this difficulty.

III. THE FISHER CONSISTENCY RESULT

The following assumptions concerning ψ will be used.

(A2) *The function* ψ: $\mathbf{R} \to \mathbf{R}$ *has the properties:*

(i) ψ is monotone nondecreasing and odd

(ii) ψ is strictly monotone on a neighborhood of zero.

(iii) ψ is continuous

Definition: A distribution function F is called *right-skewed* (RS) if $F(x)+F(-x) \leq 1$ for all x, and F is called *left-skewed* (LS) if $F(x)+F(-x) \geq 1$ for all x.

Proofs of Lemmas 1–4 below are elementary.

Lemma 1. Suppose that the random variable U has a distribution function F which gives positive probability to every neighborhood of the origin. Let ψ satisfy A2. If F is RS and $a > 0$, then $E\psi(a+U) > 0$. If F is LS and $a < 0$, then $E\psi(a+U) < 0$. If F is symmetric, then $E\psi(U) = 0$.

Lemma 2. Let X and Y be independent random variables, with the distribution of X being such that every interval has positive probability. Then the distribution of $X + Y$ gives positive probability to every interval.

Lemma 3. Let X and Y be independent random variables, with Y symmetric. If X is RS then so is $X+Y$, and if X is LS then so is $X+Y$.

Lemma 4. If U has a distribution F which is RS, then $\lambda > 0$ implies that the distribution of λU is RS and $\lambda < 0$ implies that it is LS.

The next two lemmas will also be used in order to establish Fisher consistency of $\phi(\mu)$.

Lemma 5. Let U have distribution F. For any constant $k > 0$ consider the event $M = [\,|a+U| \geq k]$, and let $F_{U|M}$ denote the distribution of U given M.

(i) If $a > 0$ and F is RS, then $F_{U|M}$ is RS.

(ii) If $a < 0$ and F is LS, then $F_{U|M}$ is LS.

(iii) If $a = 0$ and F is symmetric, then $F_{U|M}$ is symmetric.

Proof: The result (iii) is immediate, and since the arguments for (i) and (ii) are essentially the same we prove only (i). It suffices to show that for all $t \geq 0$ we have

$$P([U > t] \cap M) \;\geq\; P([U \leq -t] \cap M). \tag{3.1}$$

Note that $M = [U \geq k-a] \cup [U \leq -k-a]$, and if $a > 0$, $t \geq 0$ we have

$$P([U \geq t] \cap M) \;=\; P(U \geq t,\; U \geq k-a)$$

and

$$P([U \leq -t] \cap M) \;=\; P(U \leq -t,\; U \geq k-a) + P(U \leq -t,\; U \leq -k-a).$$

These probabilities are readily compared for two separate cases.

Case a: $k - a \leq t\,,\ t \geq 0$

Here

$$P\left([U \geq t] \cap M\right) = P(U \geq t)$$

and

$$P\left([U \leq -t] \cap M\right) \leq P(U \leq -t)$$

Since $U \sim F$ with F RS, we get (3.1).

Case b: $0 \leq t \leq k - a$

Now

$$P\left([U \geq t] \cap M\right) = P(U \geq k - a)$$

and

$$P\left([U \leq t] \cap M\right) = P\left(U \leq -k - a\right) \leq P\left(U \leq -(k - a)\right)$$

which again gives (3.1). □

Lemma 6: Let $U_1, U_2, \cdots,$ be independent and identically distributed random variables with symmetric distribution function F. Let $a_1, a_2, \cdots,$ and $h_2, h_3, \cdots,$ be constants. Let $V_1 = U_1$ and for $i = 2, 3, \cdots,$ let

$$V_i = h_i V_{i-1} + U_i\,. \tag{3.2}$$

Consider the events

$$M_i = \left[\,\left|a_i + V_i\right| \geq K_i\right], \qquad i = 1, 2, \cdots$$

where K_1 is a constant, and for each $i \geq 2$ K_i is a function of $M_1, \cdots, M_{i-1}$. Set $M^n = \bigcap_{i=1}^{n} M_i$, and let $F_{V_n |M^n}$ be the conditional distribution of V_n given M^n.

(i) If $h_2 \geq 0, \ldots, h_n \geq 0$ and $a_1 \geq 0, \ldots, a_n \geq 0$, then $F_{V_n |M^n}$ is RS.

(ii) If $h_2 \geq 0, \ldots, h_n \geq 0$ and $a_1 \leq 0, \ldots, a_n \leq 0$, then $F_{V_n |M^n}$ is LS.

(iii) If $h_2 \leq 0, \ldots, h_n \leq 0$ and $a_1 \geq 0, a_2 \leq 0, \ldots, a_n(-1)^n \leq 0$, then $F_{V_n |M^n}$ is RS or LS according if n is odd or even.

(iv) If $h_2 \leq 0, \ldots, h_n \leq 0$ and $a_1 \leq 0, a_2 \geq 0, \ldots, a_n(-1)^n \geq 0$, then $F_{V_n |M^n}$ is LS or RS according if n is odd or even.

(v) If $a_1 = a_2 = \cdots = a_n = 0$, then $F_{V_n |M^n}$ is symmetric.

Proof: The proof is by induction. For $n = 1$,

$$M_1 = \left[\,\left|a_1 + U_1\right| \geq K_1\right]$$

and so (i)–(iii) follow from Lemma 5. Now suppose the result holds for $n - 1$, and consider the case (i). Then conditioned on M^{n-1}, $h_n V_{n-1}$ is RS and U_n is symmetric. From Lemma 3 it follows that conditioned on M^{n-1}, V_n is RS. Then since K_n is fixed, when we condition on M^{n-1}, use of Lemma 5 shows that $F_{V_n |M^n}$ is RS. A similar argument yields cases (ii) to (v). □

Theorem: (Fisher Consistency) Suppose that F satisfies A1 and ψ satisfies A2. Furthermore, assume that the processes $\hat{x}_t$, s_t and x_t are jointly asymptotically stationary, and are governed by their asymptotic joint measure. If $\phi\,\phi_0 \geq 0$ and $\phi \neq \phi_0$ then for $t = 1, 2, \cdots,$

$$(\phi - \phi_0)\, E\, \hat{x}_{t-1}(\phi)\, \psi\left(\frac{r_t(\phi)}{s_t(\phi)}\right) \; < \; 0$$

where $r_t(\phi) = x_t - \phi\, \hat{x}_{t-1}(\phi)$, and

$$E\, \hat{x}_{t-1}(\phi_0)\, \psi\left(\frac{r_t(\phi_0)}{s_t(\phi)}\right) \; = \; 0\,.$$

Proof: Let $\psi_r(u) = \psi\left(\dfrac{u}{K_r^*}\right)$, where K_r^* is given by (2.14), for any fixed $r \geq 0$, consider the conditional expectation

$$E\left[\hat{x}_{t-1}(\phi)\, \psi\left(\frac{r_t(\phi)}{s_t(\phi)}\right) \,\Big|\, N_{t-1}(\phi) = t - r - 1,\; x_{t-r-1}\right]$$

$$= \; E\left[\hat{x}_{t-1}(\phi)\, \psi_r\left(r_t(\phi)\right) \big| N_{t-1}(\phi) = t - r - 1\,, \quad x_{t-r-1}\right].$$

Conditioned on $N_{t-1}(\phi) = t - r - 1$ and x_{t-r-1} we have

$$\hat{x}_{t-r-1+i}(\phi) \; = \; \phi^i\, x_{t-r-1}, \qquad i = 0, 1, \cdots, r$$

and it follows from (1.1) that

$$x_{t-r-1+i} \; = \; \phi_0^i\, x_{t-r-1} + \sum_{l=0}^{i-1} \phi_0^l\, r_{t-r-1+i-l}\,, \qquad i = 1, 2, \cdots, r+1\,.$$

Thus, conditioned on $N_{t-1}(\phi) = t - r - 1$, we have

$$r_{t-r-1+i}(\phi) \; = \; \sum_{l=0}^{i-1} \phi_0^l\, u_{t-r-1+i-l} + (\phi_0^i - \phi^i)\, x_{t-r-1}, \qquad i = 1, 2, \cdots, r+1\,.$$

Put

$$h_i \; \equiv \; \phi_0$$

$$a_i \; = \; (\phi_0^i - \phi^i)\, x_{t-r-1} \qquad i = 1, 2, \cdots, r+1$$

$$U_i \; = \; u_{t-r-1+i} \qquad i = 1, 2, \cdots, r+1\,.$$

Let V_i, $1 \leq i \leq r$ be defined by (3.2) of in Lemma 6, so that

$$V_i \; = \; \sum_{l=0}^{i-1} \phi_0^l\, u_{t-r-1+i-l}, \qquad i = 1, 2, \cdots, r+1\,.$$

and

$$r_{t-r-1+i}(\phi) \; = \; V_i + a_i\,, \qquad i = 1, 2, \cdots, r+1\,.$$

Recalling the definition of M_t^* in (2.16), let

$$M_i \; = \; M_{t-r-1+i}^*\,, \qquad i = 1, 2, \cdots, r+1$$

and note that conditioned on $N_{t-1}(\phi) = t - r - 1$ and x_{t-r-1}, we are ready to apply

Lemma 6 with $n = r+1$. We have

$$E\left[\hat{x}_{t-1}(\phi)\,\psi_r(r_t(\phi)) \;\middle|\; N_{t-1}(\phi) = t-r-1\,,\, x_{t-r-1}\right]$$
$$= \quad \phi^r x_{t-r-1}\, E\psi_r\left(V_{r+1} + a_{r+1} \;\middle|\; M^r,\, x_{t-r-1}\right). \qquad (3.3)$$

If $\phi = \phi_0$, then $a_1 = a_2 = \cdots = a_{r+1} = 0$, part (v) of Lemma 6 gives that $F_{V_r \mid M^r}$ is symmetric, and it follows from (3.2) that $F_{V_{r+1} \mid M^r}$ is symmetric as well. Then (3.3) is zero by Lemma 1.

Suppose first that $\phi_0 \in (0, 1)$. If $0 < \phi < \phi_0$ and $x_{t-r-1} > 0$, then all the a_i's are positive and $F_{V_r \mid M^r}$ is RS by Lemma 6–(i). Then $F_{V_{r+1} \mid M^r}$ is RS by Lemma 3, and Lemmas 1–2, along with A1–A2, show that (3.3) is positive. Similarly, if $\phi < \phi_0$ and $x_{t-r-1} < 0$ then the a_i are all negative, $F_{V_r \mid M^r}$ and $F_{V_{r+1} \mid M^r}$ are both LS, which gives $E\left[\psi_r\left(V_{r+1} + a_{r+1}\right) \middle| M^r\right] < 0$, and (3.3) is once again positive. Since $P(x_{t-r-1} = 0) = 0$, the result follows for $\phi \in (0, 1)$, $0 < \phi < \phi_0$. A similar argument shows that (3.3) is negative for $\phi > \phi_0$.

Now suppose that $\phi_0 \in (-1, 0)$. If $\phi < \phi_0 < 0$, $x_{t-r-1} > 0$ and r is odd, then we have $h_2 < 0, \ldots, h_r < 0$, $a_1 > 0$, $a_2 < 0, \ldots, a_r > 0$, $a_{r+1} < 0$. It follows from Lemma 6(iii) that $F_{V_r \mid M^r}$ is RS, and then by Lemmas 3–4 $F_{V_{r+1} \mid M^r}$ is LS. Hence Lemmas 1–2 and A1–A2 yield $E\left[\psi_r(V_{r+1} + a_{r+1}) \;\middle|\; M^r\right] < 0$. Since $\phi^r x_{t-r-n} < 0$, (3.3) is positive. Similar arguments show that (3.3) is positive for r even, and also for $x_{t-r-n} < 0$, r even or odd. Thus $E\left[\hat{x}_{t-1}(\phi)\,\psi(r_t(\phi)) \middle| M^r\right] < 0$, for $\phi < \phi_0 < 0$. Similar arguments show that (3.3) is negative for $\phi_0 < \phi < 0$.

If $\phi_0 = 0$, then the above arguments reveal that (3.3) is positive for $\phi < 0$ and negative for $\phi > 0$.

The result follows by averaging over the conditioning in (3.3). □

IV. CONCLUDING REMARKS

The theorem in Section 3 does not in fact give uniqueness of the root of (2.5) unless we know the sign of ϕ_0. At the present time, we have good reason to believe that the inequality of the theorem does not hold for all $\phi \in (-1, 1)$. However, in the case that (2.5) has a root may sign, we still can be Fisher consistent by choosing as an estimate the root minimizing $\sum_{t=2}^{n} \left(\dfrac{\hat{x}_t - \phi \hat{x}_{t-1}}{s_t}\right)^2$.

It would be nice to obtain Fisher consistency for the AR(p) case. Unfortunately, Fisher consistency does not hold for the p *th* order analogue ($p \geq 2$) of the hard-rejection filter-based AM-estimated treated here. It appears, however, that one or more modifications may yield Fisher consistency.

These questions will be pursued elsewhere.

REFERENCES

Bustos, O.H. (1982). "General M-estimates for contaminated pth-order autoregressive processes: consistency and asymptotic normality." *Z. Wahrsch.* **59,** 491–504.

Bustos, O.H., Fraiman, R. and Yohai, V.J. (1984). "Asymptotic behavior of the estimates based on residual autocovariances for ARMA models." In *Robust and Nonlinear Times Series Analysis,* edited by J. Franke, W. Hardle, and D. Martin, pp. 26–49. Springer, New York.

Bustos, O. and Yohai, V.J. (1986). "Robust estimates for ARMA models." *Journal of the American Statistical Association,* **81,** 155–168.

Denby, L. and Martin, R.D. (1979). "Robust estimation of the first order autoregressive parameter." *Jour. Amer. Stat. Assoc.* **74,** 140–146.

Hampel, F.R., Ronchetti, E.M., Rousseeuw, P.J. and Stahel, W.A. (1986). *Robust Statistics: The Approach Based on Influence Functions.* Wiley, New York.

Kleiner, B., Martin, R.D. and Thompson, D.J. (1979). "Robust estimation of power spectra." *Jour. Roy. Stat. Soc., Series B,* **41,** 313–351.

Kunsch, H. (1984). "Infinitesimal robustness for autoregressive processes." *The Annals of Statistics,* **12,** 843–863.

Mallows, C.L. (1980). "Some theory of nonlinear smoothers." *The Annals of Statistics* **8,** 695–715.

Martin, R.D. (1980). "Robust estimation of autoregressive models (with discussion)." In *Directions in Time Series,* edited by D.R. Brillinger and G.C. Tiao. Instit. of Math. Statistics Publication, Haywood, CA, pp. 228–254.

Martin, R.D. (1981). "Robust methods for time series." In *Applied Time Series II,* edited by D.F. Findley. Academic Press, New York, pp. 683–759.

Martin, R.D. (1982). "The Cramer-Rao bound and robust M-estimates for autoregressions." *Biometrika* **69,** 437–442.

Martin, R.D., Samarov, A. and Vandaele, W. (1983). "Robust methods for ARIMA models." In *Applied Time Series Analysis of Economic Data,* edited by A. Zellner. Econ. Res. Report ER-t, Bureau of the Census, Washington, DC.

Martin, R.D. and Thompson, D.J. (1982). "Robust resistant spectrum estimation." *IEEE Proceedings,* Vol. **70,** 1097–1115.

Martin, R.D. and Yohai, V.J. (1985). "Robustness in time series and estimating ARMA models." In *Handbook of Statistics,* **5,** edited by E.J. Hannan, P.R. Krishnaiah and M.M. Rao, pp. 119–155. Elsevier, New York.

Martin, R.D. and Yohai, V.J. (1986). "Influence functionals for time series." *The Annals of Statistics,* **14,** 781–855.

Papantoni-Kazakos, P. and Gray, R.M. (1979). "Robustness of estimators on stationary observations." *Ann. Probab.* **7,** 989–1002.

BAYES LEAST SQUARES LINEAR REGRESSION IS ASYMPTOTICALLY FULL BAYES: ESTIMATION OF SPECTRAL DENSITIES

H.D. Brunk
Department of Statistics
Oregon State University
Corvallis, Oregon 97331

ABSTRACT

The Bayes least squares linear method of estimating regression functions using orthogonal expansions may be implemented by making first an orthogonal transformation that yields uncorrelated variables. The theorem presented here gives conditions sufficient for joint asymptotic normality of the transformed variables. Thus when the coefficients are independent and normal according to the prior, one expects that in the presence of much data they will be (approximately) independent normal according to the posterior. The method is illustrated using a suggestion of Wahba (1980) to transform the problem of estimating a spectral density into a regression problem.

Key words and phrases: Bayes least squares linear; Regression; Spectral density.

I. INTRODUCTION

Bayes least squares linear (BLSL) estimators are introduced by Whittle (1957, 1958) and described explicitly and further developed by Hartigan (1969). The method was applied to estimation of coefficients of orthogonal expansions of regression functions in (Brunk, 1980). In the present paper we observe that when many observations are available we can expect the BLSL method to yield substantially the same results as a full Bayesian treatment; and we illustrate the method in the context of estimation of spectral densities. In that context, the estimators suggested will appear rather ordinary. But they are not completely ad hoc: each comes with an interpretation. And, when large samples are available, the posterior distribution of the estimator at a fixed frequency is (approximately) normal, with easily calculated standard deviation.

II. ORTHOGONAL EXPANSIONS AND THE BLSL METHOD

In order to be more explicit, we recall the description of the estimators given in Brunk (1980). Since the size of the data set is relevant here, we allow the number of

---

H.D. Brunk is with the Department of Statistics, Oregon State University, Corvallis, Oregon 97331.
This work was supported by the Office of Naval Research through N00014-81-K-0814 with Ronald Mohler as principal investigator, and the National Science Foundation through Grant MCS 80 02907-01. The author wishes to thank Richard Bucolo and Roy Rathja for their assistance; and most particularly Ronald Stillinger for most effective use of his programming expertise.

observations to appear in the notation. Thus for each integer n we have a set $\{x_{no}, x_{n1}, \cdots, x_{nn}\}$ of values of an "explanatory variable" in a space X of possible values. The regression function R_n is defined on X and is assumed to have a finite expansion in terms of specified functions r_{no}, R_{no} and $\{\phi_{nr}, r = 0,1,...,n\}$:

$$R_n(x) = R_{no}(x) + r_{no}(x)\sum_{r=0}^{n}\beta_{nr}\phi_{nr}(x), \qquad x \in X. \tag{2.1}$$

The observations or responses $(Y_{no}, Y_{n1},...,Y_{nn})$ are assumed independent, and Y_{nj} is assumed to have mean

$$E(Y_{nj}) = R_n(x_{nj}) \tag{2.2}$$

and variance

$$\mathrm{Var}(Y_{nj}) = 1/\lambda_n \pi_{nj}, j = 0,1,...,n, \tag{2.3}$$

where $\pi_{no},...,\pi_{nn}$ are known. We shall assume also that λ_n is known, though in practice λ_n may be estimated from the data.

(One can state less restrictive assumptions that suffice. In what follows, as elsewhere, a tilde underline indicates a random entity, and ":=" is used between two expressions when the left is defined by the right. These assumptions are that the linear expectation of Y_{nj} be $R_n(x_{nj})$:

$$LE(Y_{nj} \mid \underset{\sim n}{\beta} = \beta_n) = R_n(x_{nj}), j = 0,1,...,n,$$

where $\beta_n := (\beta_{no},...,\beta_{nn})'$; and the linear covariance matrix of $Y_n := (Y_{no},...,Y_{nn})'$ given $\beta_n = \beta_n$ have entries

$$E([Y_{ni} - R_n(x_{ni})][Y_{nj} - R_n(x_{nj})] \mid \underset{\sim n}{\beta} = \beta_n) = \delta_{ij}/\lambda_n \pi_{nj}, \qquad i,j = 0,1,...,n;$$

here δ_{ij} is the Kronecker delta: $\delta_{ij} = 1$ if $i = j, \delta_{ij} = 0$ if $i \neq j$.)

The functions $\{\phi_{nr}, r = 0,1,...,n\}$ are assumed selected so as to be orthonormal with respect to the design of the experiment:

$$\sum_{j=0}^{n} \pi_{nj} r_{no}^2(x_{nj})\phi_{nr}(x_{nj})\phi_{ns}(x_{nj}) = \delta_{rs}. \tag{2.4}$$

The function R_{no} is thought of as a prior mean, so that one sets

$$E(\underset{\sim nr}{\beta}) = 0, \; r = 0,1,...,n. \tag{2.5}$$

It is argued in Brunk (1980) that since each coefficient β_{nr} has an interpretation independent of β_{ns} for $s \neq r$, it may often be reasonable to assign $\underset{\sim n}{\beta} := (\underset{\sim ns}{\beta}, \underset{\sim ni}{\beta}, \ldots, \underset{\sim nn}{\beta})'$ a joint prior distribution according to which the components $\underset{\sim no}{\beta}, \ldots, \underset{\sim nn}{\beta}$ are independent; and we set

$$\tau_{nr} := 1/\mathrm{Var}(\underset{\sim nr}{\beta}), \; r = 0,1,...,n. \tag{2.6}$$

Set

$$\begin{aligned} U_{nr} &:= \textstyle\sum_{j=0}^{n} \pi_{nj} r_{no}(x_{nj})\phi_{nr}(x_{nj})[Y_{nj} - R_{no}(x_{nj})] \\ &= \textstyle\sum_{j=0}^{n} c_{nrj} W_{nj}, \; r = 0,1,...,n, \end{aligned} \tag{2.7}$$

where

$$c_{nrj} := \sqrt{\pi_{nj}}\, r_{no}(x_{nj})\phi_{nr}(x_{nj}) \tag{2.8}$$

and

$$W_{nj} := \sqrt{\pi_{nj}}[Y_{nj} - R_{no}(x_{rj})], \qquad r,j = 0,\ldots,n. \tag{2.9}$$

That is, the random vector

$$U_n := (U_{no}, U_{nr}, \ldots, U_{nn})'$$

is obtained by applying the orthogonal transformation C_n to the vector $W_n := (W_{no}, \ldots, W_{nn})'$, where

$$(C_n)_{rj} := c_{nrj}, \; r = 0,1,\ldots,n, \; j = 0,1,\ldots,n.$$

It follows from (2.4) that

$$E(U_{nr} \mid \underset{\sim n}{\beta} = \beta_n) = \beta_{nr} \tag{2.10}$$

and that

$$[cov(U_{nr}, U_{ns} \mid \underset{\sim n}{\beta} = \beta_n] = \delta_{rs}/\lambda_n. \tag{2.11}$$

Then the linear expectation of $\underset{\sim n}{\beta}$ given $U_{no}, \ldots, U_{nn}$ is given by

$$\underset{\sim nr}{\hat{\beta}} = \lambda_n U_{nr}/(\lambda_n + \tau_{nr}), \; r = 0,1,\ldots,n \tag{2.12}$$

and the linear variances and covariances are given by

$$E(\underset{\sim nr}{\beta} - \underset{\sim nr}{\hat{\beta}})^2 = 1/(\lambda_n + \tau_{nr}), r = 0,1,\ldots,n, \tag{2.13}$$

$$E(\underset{\sim nr}{\beta} - \underset{\sim nr}{\hat{\beta}})(\underset{\sim ns}{\beta} - \underset{\sim ns}{\hat{\beta}}) = 0 \text{ for } r \neq s. \tag{2.14}$$

For fixed x, the linear expectation of $R_n(x)$ is

$$\hat{R}_n(x) = R_{no}(x) + r_{no}(x) \sum_{r=0}^{n} \underset{\sim nr}{\hat{\beta}} \phi_{nr}(x) \tag{2.15}$$

and its linear variance is

$$E([R_n(x) - \hat{R}_n(x)]^2) = r_{no}^2(x) \sum_{r=0}^{n} \phi_{nr}^2(x)/(\lambda_n + \tau_{nr}). \tag{2.16}$$

Note that the method does not, in general, provide posterior covariances.

It will be useful to note that when $R_{no} = 0$ and $\tau_{nj} = 0$ for j=0,1,...,n, the formula for $\hat{R}_n(x)$ provides the ordinary least squares regression of Y on x with weights τ_{nj}, j=0,1,...,n . Indeed, when k is fixed, $k <= n$, the orthogonality properties of the functions $\{\phi_{nr} := r = 0,1,\ldots,n\}$ lead to $\sum_{r=0}^{k} \hat{a}_{nr} \phi_{nr}$ as ordinary least squares estimator of R_n, where

$$\hat{a}_{nr} := \sum_{j=0}^{n} \pi_{nj} \phi_{nr}(x_{nj}) Y_{nj}, r = 0,1,\ldots,k.$$

Note that $\hat{a}_{nr} = U_{nr}$ if $R_{no} = 0$ and $\hat{a}_{nr} = \hat{\beta}_{nr}$ if also $\tau_{nr} = 0$, r=0,1,...,k.

Of course, the BLSL estimators may be considered from a conventional point of view. That is, one may choose, as is customary when estimating a regression function R_n, some family of functions considered adequate for representing it. One may then orthogonalize these functions with respect to the design, to obtain functions $\{\phi_{nr}, r = 0,1,\ldots,n\}$; and then specify a function R_{no} and "precisions" $\{\tau_{nr}' \; r = 0,1,2,\ldots,n\}$, thus finally obtaining an estimator of the regression function R_n. But such an estimator is not completely ad hoc; it comes with an interpretation. One realizes that one is considering the same estimator that another investigator would use who was applying the Bayes least squares linear method, specifying R_{no} as a prior mean, and the $\{\tau_{nr}, r = 0,1,...,n\}$ as precisions of the coefficients in the expansion of R_n. And one may like to bear that in mind when specifying R_{no} and the precisions.

In principle, an investigator who is well acquainted with the functions $\{\phi_{nr}\}$ to be used, and who also has a clear and definite opinion as to the probable shape of the regression function to be estimated, can specify, more or less uniquely, a prior mean and prior precisions. But in practice, there may be a rather wide variety of specifications that all seem reasonable. One may then examine a number of estimates arising from a range of "reasonable" priors. As to the specification of the prior mean, R_{no}, a heuristic argument is given in Brunk (1981) that it is often reasonable to fit the data -- roughly -- by a member of a family of smooth functions depending on only one or two parameters. That leaves still the precisions, $\tau_{nr}, r = 0,1,...,n$. Two suggestions come from thinking of them as reciprocals of prior variances of the parameters $\underset{\sim}{\beta}_{nr}, r = 0,1,2,...,n$.

(i) If the functions ϕ_{nr} oscillate more and more rapidly with increasing r, one can express an opinion that the estimate $\hat{R}_n$ is "smooth" by specifying large values of τ_{nr} when r is large.

(ii) The precisions should be specified independently of the data.

III. APPROXIMATE NORMALITY

Now let us briefly imagine that the random variables $U_{no},...,U_{nn}$ were observed (rather than $Y_{no}, \ldots, Y_{nn}$), that they were independent according to their joint distributions given $\underset{\sim}{\beta}_n$, and that

$$U_{nr} \mid \underset{\sim}{\beta}_n = \beta_n \sim N(\beta_{nr}, 1/\lambda_n). \tag{3.1}$$

Suppose also that β_n is given a joint prior distribution according to which its components $\underset{\sim}{\beta}_{no}, \ldots, \beta_{nn}$ are independent, and

$$\underset{\sim}{\beta}_{nr} \sim N(0, 1/\tau_{nr}),\ r = 0,1,...,n. \tag{3.2}$$

Then these components are also independent and normally distributed according to their posterior distribution, with

$$E(\underset{\sim}{\beta}_{nr} | U_{no} = u_{no}, \ldots, U_{nn} = u_{nn}) = \lambda_n u_{nr}/(\lambda_n + \tau_{nr}), \tag{3.3}$$

(cf. (2.12)) and

$$Var(\underset{\sim}{\beta}_{nr} | U_{no} = u_{no}, \ldots, U_{nn} = u_{nn}) = 1/(\lambda_n + \tau_{nr}),\ r = 0,1,...,n \tag{3.4}$$

(cf. (2.13), (2.14)).

We are interested particularly in contexts in which one expresses an opinion as to the "smoothness" of R_n by assigning large precisions τ_{nr} to the coefficients of rapidly varying functions ϕ_{nr} in the expansion of R_n. Then, typically, there is a positive integer m such that the posterior mean and variance of $\underset{\sim}{\beta}_{nr}$ are so near zero for $r > m$ that corresponding terms in the expansion can be neglected; and this is so also for the posterior linear expectation and for $E(\hat{\underset{\sim}{\beta}}_{nr} - \underset{\sim}{\beta}_{nr})^2$ in the BLSL method.

When the observations $Y_{no},...,Y_{nn}$ are not jointly normal, but n is large, we shall argue that often $U_{no},...,U_{nn}$ will have approximately a joint normal distribution (cf. Theorem 4.1, to follow). The data $U_{no},...,U_{nn}$ are fully equivalent to the original data $Y_{no},...,Y_{nn}$: each may be obtained from the other by an orthogonal linear transformation. And for some positive integer m, $U_{n,m+1}, \ldots, U_{nn}$ may safely be ignored, so that if $\underset{\sim no}{\beta}, \ldots, \underset{\sim nn}{\beta}$ are given the multinormal prior distribution described above, then according to their joint posterior distribution they are (approximately) jointly normally distributed with posterior means given by (2.12) and (3.3) and posterior variances by (2.13) and (3.4). A theorem (Theorem 4.1) that suggests this approximation is given in Section 4, and its proof in the appendix.

IV. ASYMPTOTIC NORMALITY

We consider triangular arrays v : $V_{no}, V_{ni},..., V_{n,k_n}$, n=1,2,..., where $k_n \rightarrow \infty$ as $n \rightarrow \infty$, and where $E(V_{nj}) = 0$, $Var(V_{nj}) = 1$, $j = 0,1,...,k_n$. We shall argue that this array is asymptotically normal, given $\underset{\sim n}{\beta} = \beta_n$, when

$$V_{nr} := [U_{nr} - E(U_{nr})]/[Var(U_{nr}]^{1/2}, \qquad r = 0,1,...,k_n \tag{4.1}$$

provided that k_n does not grow too fast. We use Mallows's (1972) definition of asymptotic normality: the array v is jointly asymptotically normal (j.a.n.) if for every array, A, of reals $a_{no}, a_{nj},..., a_{n,kn}$, $n = 0,1,...$, such that $\sum_{r=0}^{k_n} a_{nr}^2 = 1$, the random variable $\sum_{r=0}^{k_n} a_{nr} V_{nr}$ converges in distribution to the standard normal distribution. (Mallows (1972) observes that this implies that for each d, $(V_{no}, \ldots, V_{nd})$ converges in distribution to the standard d-dimensional normal distribution.)

Theorem 4.1. Let there be a positive number M such that

$$(\lambda_n \pi_{nj})^{3/2} E\, |Y_{nj} - \mu_{nj}|^3 <= M, \tag{4.2}$$

where

$$\mu_{nj} := E(Y_{nj}) = R_n(x_{nj}), \quad j = 0,1,...,n, \quad n = 1,2,... \tag{4.3}$$

And suppose that

$$\max\{\sqrt{\pi_{nj}}\, |r_{no}(x_{nj})| \textstyle\sum_{r=0}^{k_n} |\phi_{nr}(x_{nj})| : j = 0,1,...,n\} \rightarrow 0 \quad \text{as } n \rightarrow \infty. \tag{4.4}$$

Then the array v is j.a.n.

The proof, given in the appendix, consists of showing in a straightforward way that the characteristic function of $\sum_{r=0}^{k_n} a_{nr} V_{nr}$, evaluated at a real number t, converges to $\exp(-t^2/2)$ as $n \rightarrow \infty$. This theorem presents an instance of a phenomenon studied by Mallows (1969). Note that the random variables U_{nr} are obtained via an orthogonal transformation from the random variables W_{nj}; and that while these are independent, they need not be normally distributed. Although the inverse orthogonal transformation would recover the original non-normal random variables, the random variables U_{nr} are nevertheless j.a.n. (Mallows (1969) proves a theorem with a somewhat stronger conclusion, described in terms of the integrated squared difference between the standard normal distribution function and the distribution function it approximates. As stated, Mallows's theorem requires independent, identically distributed random variables. While the method of proof appears to allow a relaxation of that requirement, it does seem to require that the distributions not be too nearly of lattice type.)

V. BLSL ESTIMATION OF SPECTRAL DENSITIES

We consider the problem of estimating the spectral density of a stationary, purely nondeterministic time series

$$X_t = \sum_{s=0}^{\infty} \alpha_s \underset{\sim}{\epsilon}_{t-s}$$

(not necessarily Gaussian), where the random variables $\underset{\sim}{\epsilon}_t$, $t = \ldots,-1,0,1, \cdots$ are independent with zero means and common variance. The coefficients α_s, $s = 0,1,\ldots$, are unknown, but we assume that $\sum_{t=0}^{\infty} |\alpha_t| < \infty$ and that

$$\sup_t E\left[\underset{\sim}{\epsilon}_t^2 1_{\{|\underset{\sim}{\epsilon}_t| > c\}}\right] \to 0 \text{ as } c \to \infty$$

(cf. Anderson (1971), page 482). The spectral density, to be estimated, is

$$f(\omega) = \sum_{t=-\infty}^{\infty} r(t) \exp(2\pi i \omega t), \; -1/2 <= \omega <= 1/2,$$

where

$$r(t) := E(X_s X_{s+t})$$

is the covariance function. We follow Wahba (1980) in formulating the problem as an ordinary regression problem, with values of the log periodogram (cepstrum) as data. Let $X_1, X_2, \ldots, X_{2n}$ be observed. We have

$$I_n(\omega) := (1/2n) |\sum_{t=1}^{2n} X_t \exp(2\pi i \omega t)|^2, \; -1/2 <= \omega <= 1/2,$$

$$I_{nj} := I_n(j/2n) \; (= I_n(-j/2n))$$

$$= f(j/2n) T_{nj},$$

where

$$T_{nj} := I_{nj} / f(j/2n), \; j = 0,1,2,\ldots,n.$$

Asymptotically, as $n \to \infty$, the random variables $\{T_{nj},\; j=0,1,\ldots,n\}$ are independent, with T_{no} and T_{nn} distributed as $\chi^2(1)$ and $2T_{nj} \sim \chi^2(2)$ for $j=1,2,\ldots,n-1$ (Anderson, 1971, pp. 484-485). We set

$$Y_{nj} := \log I_{nj} + C_j, \; j = 0,1,\ldots,n,$$

where $C_o = C_n := (ln 2+\gamma), C_j := \gamma$, $j=1,2,\ldots,n-1$, and where γ is the Euler-Mascheroni constant, approximately 0.57721. Then the random variables $\{Y_{nj},\; j=0,1,\ldots,n\}$ are asymptotically independent, with $\{Y_{nj},\; j=1,2,\ldots,n-1\}$ asymptotically identically distributed.

We set

$$R_n(x) := \log f(x/2) \text{ for } 0 <= x <= 1.$$

According to the asymptotic distribution,

$$E(Y_{nj}) = R_n(j/n) = R_n(x_{nj})$$

where $x_{nj} := j/n$ for $j=0,1,2,\ldots,n$, while

$$\mathrm{Var}(Y_{nj}) = \pi^2/6, j=1,2,\ldots,n-1,$$
$$\mathrm{Var}(Y_{nj}) = \pi^2/2 \text{ for } j=0,n.$$

We shall conduct the analysis and carry out the computations as if $Var(Y_{nj}) = \pi^2/3$ for $j=0,n$, since in any case the influence of these two terms is negligible for large n. In the notation of Section 2, then, we take

$$\pi_{no} := \pi_{nn} := 1/2,\ \pi_{nj} := 1 \text{ for } j = 1,2,\ldots,n-1,$$

and (5.1)

$$\lambda_n := 6/\pi^2,\ n = 1,2,\ldots\ .$$

The functions ϕ_{nr} are given by

$$\phi_{no}(x) := 1/\sqrt{n}\ ,$$
$$\phi_{nr}(x) := (2/n)^{1/2}\cos \pi r x,\ r = 1,2,\ldots,n, x \epsilon [0,1].$$

The function r_{no} used is identically 1. Note that these functions ϕ_{nr} are orthonormal with respect to the design, as is required by (2.4).

It is most convenient to take as R_o a linear combination of the functions $\{\phi_{nr} := r = 0,1,\ldots,n\}$. As we noted earlier, a heuristic argument is given in Brunk (1981, page 117) that it is reasonable for the investigator to select as prior mean R_o a regression function that is consistent with both the data and his opinion as to its shape. In particular, the investigator may simply use ordinary least squares with weights $\tau_{no}, \ldots, \pi_{nn}$ to choose coefficients $a_o, a_1, \ldots, a_k$ for $k = 1$ *or* 2 *or* 3 in fitting the function $\sum_{r=0}^{k} a_r \phi_{nr}$ to the data. This would yield

$$R_o(x) = \sum_{r=0}^{k} a_r \phi_{nr}(x)$$

where

$$a_r := \sum_{j=0}^{n} \pi_{nj} \phi_{nr}(x_{nj}) Y_{nj}.$$

And then (cf. the end of Section 2) if $\tau_0 = \tau_1 = \cdots = \tau_k = 0$, we have

$$\hat{R}_n(x) = \sum_{r=0}^{n} \hat{\beta}_{nr}' \phi_{nr}(x), \text{ where}$$
$$\hat{\beta}_{nr}' := \lambda_n U_{nr}' / (\lambda_n + \tau_{nr}),\quad r = 0,1,\ldots,n\ ,$$

and

$$U_{nr}' := \sum_{j=0}^{n} \pi_{nj} \phi_{nr}(x_{nj}) Y_{nj},\quad r = 0,1,\ldots,n\ .$$

In other terms, if the investigator chooses to consider a specification of prior mean and precisions that has R_0 as the ordinary least squares estimator of R_n as a linear function of $\phi_{no}, \phi_{n1}, \ldots, \phi_{nk}$, and has prior precisions $\tau_{no} = \tau_{n1} = \cdots \tau_{nk} = 0$, then $\hat{R}_n$ is precisely what it would be if he took $R_o \equiv 0$ (in which case U_{nr} would become U_{nr}').

The functions ϕ_{nr} described above depend -- in a simple way -- on n. It is more convenient when considering specification of precisions to use functions independent of n: $\phi_r^* := (n/2)^{1/2} \phi_{nr}$, so that

$$\phi_o^* := \sqrt{1/2},\ \phi_r^*(x) := \cos \pi r x,\ r > 0 \tag{5.2}$$

We set

$$\beta_{nr}^* := (2/n)^{1/2} \beta_{nr},\quad r = 0,1,\ldots,n.$$

Then -- n being fixed -- we are assuming that R_n has an expansion

$$R_n(x)=R_{no}(x) + \sum_{r=0}^{n}\beta_{nr}^{*}\,\phi_r^{*}(x), \quad x \in [0,1].$$

While formally the assumed expansion of R_n depends on n, we consider that terms of large index are negligible; and each term is in fact independent of n, so that we write

$$R(x) = R_o(x) + \sum_{r=0}^{n}\beta_r^{*}\,\phi_r^{*}(x), \quad x \in [0,1]. \tag{5.3}$$

The coefficients $\{\beta_r^{*}, r = 0,1,...,n\}$, modeled as random variables, have means

$$E(\beta_r^{*}) = 0$$

and precisions

$$\tau_r^{*}:=1/Var(\beta_r^{*}).$$

In the present context and notation, Equation (2.12) becomes

$$\hat{\beta}_{nr}^{*} = U_r^{*}/(n/2 + \pi^2\tau_r^{*}/6), \tag{5.4}$$

where

$$U_r^{*} := \sum_{j,o}^{n} \pi_{nj}\phi_r^{*}(j/n)[Y_{nj}-R_o(j/n)], \quad r = 0,1,...,n. \tag{5.5}$$

We note that $E(2U_r^{*}/n\,|\beta^{*}) = \beta_r^{*}$, and that $2U_r^{*}/n$ is the ordinary least squares estimate of β_r^{*} with weights π_{ni}, $i = 0,1,...,n$. For fixed $x \in [0,1]$, the posterior linear expectation of $R(x)$ is

$$\hat{R}_n(x) = R_o(x)+\sum_{r=0}^{n}\underset{\sim nr}{\beta}^{*}\,\phi_r^{*}(x) \tag{5.6}$$

and its linear variance is

$$E[R(x) - \hat{R}_n(x)]^2 = \sum_{r=0}^{n}[\phi_r^{*}(x)]^2/(3n/\pi^2 + \tau_r^{*}). \tag{5.7}$$

In view of Theorem 4.1, when n is large we expect the posterior distribution of $R(x)$ to be approximately normal with mean $\hat{R}_n(x)$ given by (5.6) and variance given by (5.7).

VI. EXAMPLE

We use Wahba's example (1980):

$$X_t = \sum_{k=1}^{3} \gamma_k\, X_{t-k} + \epsilon_t,$$

where $\gamma_1 = 1.4256$, $\gamma_2 = -0.7344$, $\gamma_3 = 0.1296$, and where the random variables $\underset{\sim t}{\epsilon}$, $t=...,-2,-1,0,1,2,\cdots$ are independent, each having the standard normal distribution. (Of course this does not imply that the random variables Y_{nj} are normally distributed. We note that the purpose of this example is simply to illustrate the method described in Section 5. We have not attempted simulations intended to discover how nearly normal the regression estimators are for various values of n.) The simulation was carried out starting with $X_{-30} = 0$ and then discarding X_{-30}, X_{-29},...,X_0. (These observed X_t are not to be confused with $x_{nj} := j/n$ in the formulas in Section 5.) One set of 256 points was obtained in this way ($n = 128$), as also a larger set of 1024 ($n = 512$) containing the first.

The function R_o used is defined by:

$$R_o(x) := 0.1 + 2.9\cos(\pi x) + 0.5\cos(2\pi x), \quad 0 \leq x \leq 1.$$

The accompanying Tables 1 and 2, and Figures 1 through 14, relate to the following four specifications of precisions

A: $\tau_r^* := (0.21r)^8$,

B: $\tau_r^* := (0.00024)(6.4)^r$,

C: $\tau_r^* := 0.004(4)^r$,

and

D: $\tau_r^* := 0.1(3)^r$,

Table 1 indicates the "damping" effect of each specification of precisions; that is, the entry in the table is $1/(1+\pi^2\tau_r^*/3n)$, the factor by which the ordinary least squares estimate, $2U_r^*/n$, is multiplied to obtain $\hat{\beta}_r^*$, when $n = 128$ (256 observations). The entries in Table 2 are for $n = 512$ (1024 observations). Each specification of $\{\tau_r^*,\ r = 0,1,\ldots,\}$ leads to a "window estimator" that could be considered from a conventional point of view. Any or all of these specifications might appear reasonable to an investigator. The precisions specified under D increase most rapidly, and might be expected to lead to the smoothest estimates of R. Initially, the precisions A increase somewhat more rapidly than those of B, though eventually those of B increase much more rapidly. In fact, those of B were deliberately selected (by regression of log τ_r^* on r, from A) so as to be near those of A for $r \leq 10$.

Figure 1 shows both the true $R(x) := \log f(x/2)$ and the estimate $\hat{R}$ obtained from the 256 observations ($n = 128$) using precisions A. Figures 2, 3, and 4, show estimates obtained through precisions B, C and D, respectively. Figure 5 shows first and fourth estimates, together with the true R. Figure 6 provides the graphs for the first prior, but based on the 1024 observations ($n = 512$), and Figure 7 is for the fourth prior. Figure 8 shows first and fourth together.

Figure 9 shows the spectral density estimate and the true spectral density, graphed against twice the frequency, for the first prior (A), and 256 observations. The curves lying above and below the graph of the estimate indicate the precision of the estimate in the following way. For fixed x, the asymptotic theory leads us to expect $\hat{R}(x)$ to be approximately normally distributed according to its posterior distribution. Its (approximate) posterior variance is given by (5.7). Letting $\sigma(x)$ denote the square root of this posterior variance, the upper and lower graphs are graphs of $\exp[\hat{R}(x) \pm \sigma(x)]$. Figure 10 gives the same information for the fourth prior (D), and Figure 11 shows first and fourth estimates together. Figures 12 and 13 compare with Figures 9 and 10, but for the case of 1024 observations. Spectral density estimates for first and fourth priors are shown together in Figure 14, for the case of 1024 observations.

REFERENCES

1. Anderson, T.W. (1971). *The Statistical Analysis of Time Series.* John Wiley and Sons, Inc.

2. Brunk, H.D. (1980). "Bayesian least squares estimates of univariate regression functions," *Comm. Statist. -Theor. Meth.,* A9(11), 1101-1136.

3. Brunk, H.D. (1981). "Estimation of stimulus-response curves by Bayesian least squares," *Psychometrika* 46, 115-128.

4. Chung, K.L. (1968). *A Course in Probability Theory,* Harcourt, Brace, and World.

5. Hartigan, J.A. (1969). "Linear Bayes Methods," *J. Royal Statist. Soc. B,* 31, 446-454.

6. Mallows, C.L. (1969). "Joint normality induced by orthogonal transformations," Bell Telephone Laboratories Memorandum.

7. Mallows, C.L. (1972). "Linear processes are nearly Gaussian," *J. Appl. Prob. 4,* 313-329.

8. Wahba, G. (1980). "Automatic smoothing of the log periodogram," *J. Amer. Statist. Ass'n. 75,* 122-132.

9. Whittle, P. (1957). "Curve and periodogram smoothing," *J.Royal Statist. Soc. B,* 19, 38-46.

10. Whittle, P. (1958). "On the smoothing of probability density functions," *J. Royal Statist. Soc. B,* 20, 334-343.

APPENDIX

Proof of Theorem 4.1

Set

$$Z_{nj} := [Y_{nj} - R_n(x_{nj})]; \tag{A1}$$

then $E(Z_{nj}) = 0$, $Var(Z_{nj}) = 1$, for j=0,1,...,n ; and $Z_{no}, \ldots, Z_{nn}$ are independent. From (2.7) we have

$$U_{nr} = \sum_{j=0}^{n} \pi_{nj} \, r_{no}(x_{nj}) \, \phi_{nr}(x_{nj})[Z_{nj}/\sqrt{\lambda_n \, \pi_{nj}}) + R_n(x_{nj}) - R_{no}(x_{nj})],$$

and from (2.1) and (2.4) we have

$$U_{nr} = (1/\sqrt{\lambda_n})\sum_{j=0}^{n}\sqrt{\pi_{nj}}\, r_{no}(x_{nj})\phi_{nr}(x_{nj})Z_{nj} + \beta_{nr}, \; r = 0,\ldots,n.$$

Then from (2.10), (2.11) and (4.1),

$$V_{nr} = \sum_{j=0}^{n}\sqrt{\pi_{nj}}\,\phi_{nr}(x_{nj})r_{no}(x_{nj})Z_{nj}, \; r = 0,1,\ldots,n. \tag{A2}$$

Let f_{nj} be the characteristic function of Z_{nj};

$$f_{nj}(t) := E[\exp(itZ_{nj})], \; j = 0,1,\ldots,n. \tag{A3}$$

Let k_n be an increasing sequence of integers satisfying the hypotheses of Theorem 4.1, and let $\{a_{nj}, j = 0,1,\ldots,k_n, \; n = 0,1,\ldots\}$ be an array of real numbers such that $\sum_{j=0}^{k_n} a_{nj}^2 = 1$. Since $Z_{no}, \ldots, Z_{nn}$ are independent, it follows from (A2) that the characteristic function of $\sum_{r=0}^{k_n} a_{nr} \, V_{nr}$,

$$f_n^*(t) := E[\exp(it\sum_{r=0}^{k_n} a_{nr} \, V_{nr})], \tag{A4}$$

is given by

$$f_n^*(t) = \Pi_{j=0}^n f_{nj}([\sqrt{\pi_{nj}}\, r_{no}(x_{nj}) \sum_{r=0}^{k_n} a_{nr}\phi_{nr}(x_{nj})]\ t).$$

For fixed t, set

$$t_{nj} := [\sqrt{\pi_{nj}}\ r_{no}\ (x_{nj}) \sum_{r=0}^{k_n} a_{nr}\phi_{nr}(x_{nj})]\, t\,; \tag{A5}$$

then

$$f_n^*(t) = \Pi_{j=0}^n f_{nj}(t_{nj}). \tag{A6}$$

Since $E(Z_{nj}) = 0$ and $Var(Z_{nj}) = 1$, $j = 0,1,\dots,n$,

$$f_{nj}(t) = 1 - t^2/2 + (\alpha_{nj}/6)|t|^\beta E\,|Z_{nj}|^\beta,$$

where $|\alpha_{nj}| <= 1$, and

$$f_{nj}(t_{nj}) = 1 + \theta_{nj}, \tag{A7}$$

where

$$\theta_{nj} := -t_{nj}^2/2 + (\alpha_{nj}/6)|t_{nj}|^\beta E\,|Z_{nj}|^\beta. \tag{A8}$$

Since $|a_{nr}| <= 1$, r $= 0,1,\dots,\ k_n$, by Hypothesis (4.3), we have

$$\max\{|t_{nj}| : j = 0,1,\dots,n\} \to 0 \text{ as } n \to \infty, \tag{A9}$$

for fixed $t \in \mathbb{R}$. And according to (4.2), $E\,|Z_{nj}|^\beta <= M$ for all j and n , so that

$$\max\{|\theta_{nj}| : j = 0,1,\dots,n\} \to 0 \text{ as } n \to \infty. \tag{A10}$$

From (2.4) and (A5),

$$\begin{aligned}\sum_{j=0}^n t_{nj}^2 &= t^2 \sum_{j=0}^n \pi_{nj}\ r_{no}^2(x_{nj}) \sum_{r=0}^{k_n} a_{nr}\ \phi_{nr}(x_{nj}) \sum_{s=0}^{k_n} a_{ns}\ \phi_{ns}(x_{nj}) \\ &= t^2 \sum_{r=0}^{k_n} a_{nr}^2,\end{aligned}$$

so that

$$\sum_{j=0}^n t_{nj}^2 = t^2. \tag{A11}$$

Then

$$\begin{aligned}\sum_{j=0}^n |\theta_{nj}| &<= t^2/2 + (M/6) \sum_{j=0}^n |t_{nj}|^\beta \\ &<= t^2/2 + (Mt^2/6)\max\{|t_{nj}| : j = 0,1,\dots,n\},\end{aligned}$$

so that by (A9), for fixed t,

$$\sum_{j=0}^n |\theta_{nj}| \text{ is bounded.} \tag{A12}$$

Again, from (A8) and (A11),

$$|\sum_{j=0}^n \theta_{nj} + t^2/2| <= (Mt^2/6)\max\{|t_{nj}| : j = 0,1,\dots,n\},$$

so that by (A9)

$$\sum_{j=0}^n \theta_{nj} \to -t^2/2 \text{ as } n \to \infty.$$

It follows from (A6), (A7), (A10), (A12), and (A13) that

$$f_n^*(t) = \Pi_{j=0}^n f_{nj}(t_{nj}) = \Pi_{j=0}^n (1 + \theta_{nj}) \to \exp(-t^2/2)$$

as $n \to \infty$ (Chung, 1968, page 184), for each real t. So $\sum_{r=0}^{k_n} a_{nr} V_{nr}$ converges in law to the standard normal distribution, and the array v is jointly asymptotically normal.

TABLE 1

MULTIPLIERS FOR n = 128

r	A	B	C	D
0	1.000	1.000	1.000	0.997
1	1.000	1.000	1.000	0.992
2	1.000	1.000	0.998	0.977
3	0.999	1.000	0.993	0.935
4	0.994	0.999	0.974	0.828
5	0.963	0.993	0.905	0.616
6	0.860	0.959	0.704	0.348
7	0.641	0.787	0.373	0.151
8	0.380	0.365	0.129	0.056
9	0.193	0.083	0.036	0.019
10	0.093	0.014	0.009	0.007
11	0.046	0.002	0.002	0.002
12	0.023	0.000	0.001	0.001
13	0.012	0.000	0.000	0.000
14	0.007	0.000	0.000	0.000
15	0.004	0.000	0.000	0.000
16	0.002	0.000	0.000	0.000
17	0.001	0.000	0.000	0.000
18	0.001	0.000	0.000	0.000
19	0.001	0.000	0.000	0.000
20	0.000	0.000	0.000	0.000

TABLE 2

MULTIPLIERS FOR n = 512

r	A	B	C	D
0	1.000	1.000	1.000	0.999
1	1.000	1.000	1.000	0.998
2	1.000	1.000	1.000	0.994
3	1.000	1.000	0.998	0.983
4	0.998	1.000	0.993	0.951
5	0.991	0.998	0.974	0.865
6	0.961	0.990	0.905	0.681
7	0.877	0.936	0.704	0.416
8	0.710	0.697	0.373	0.192
9	0.489	0.265	0.129	0.073
10	0.292	0.053	0.036	0.026
11	0.161	0.009	0.009	0.009
12	0.087	0.001	0.002	0.003
13	0.048	0.000	0.001	0.001
14	0.027			
15	0.016			
16	0.009			
17	0.006			
18	0.004			
19	0.002			
20	0.002			

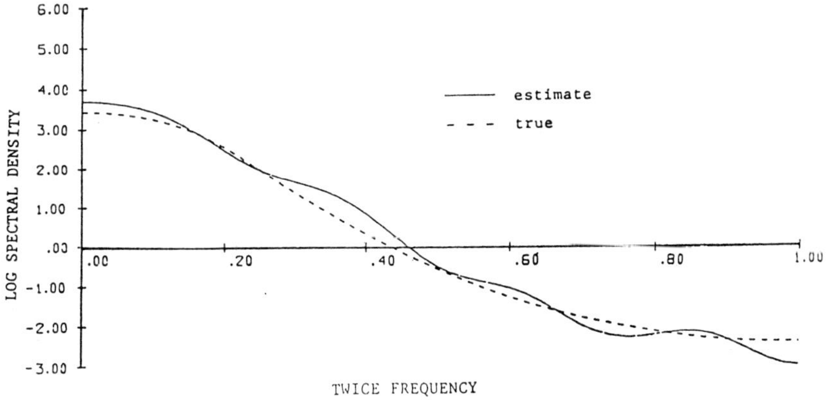

Figure 1. 256 points, first prior

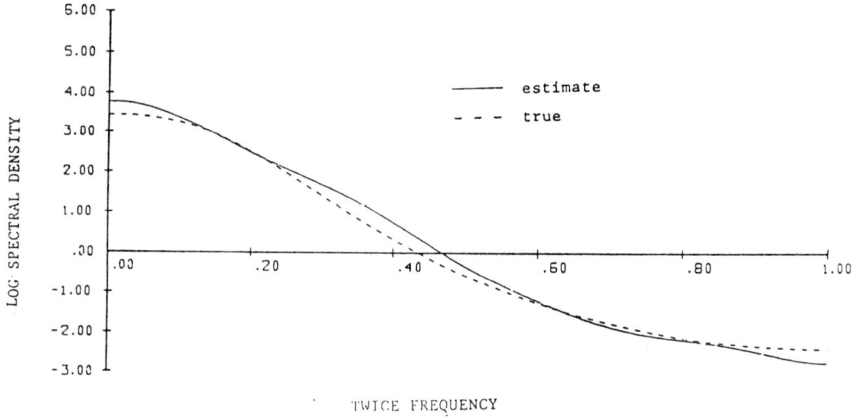

Figure 2. 256 points, second prior

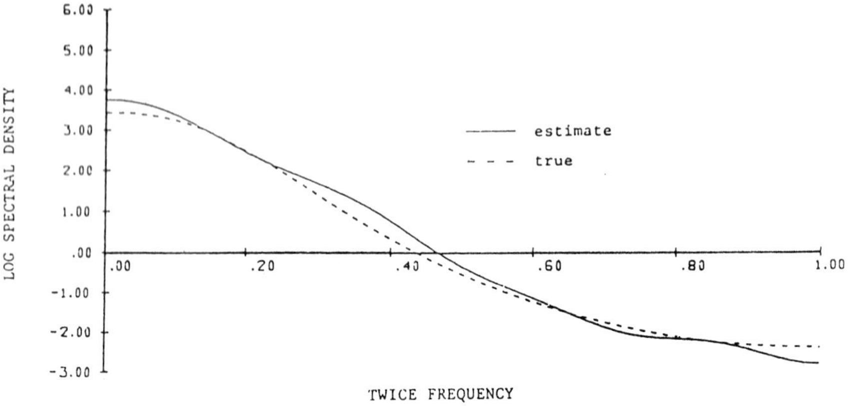

Figure 3. Third prior

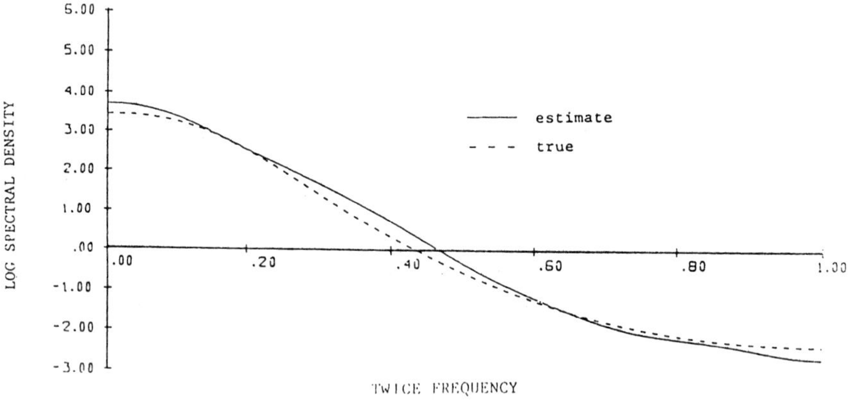

Figure 4. Four prior

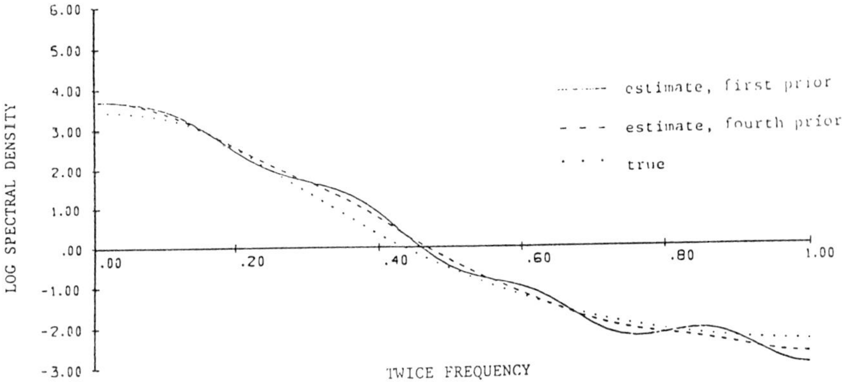

Figure 5. First and fourth priors.

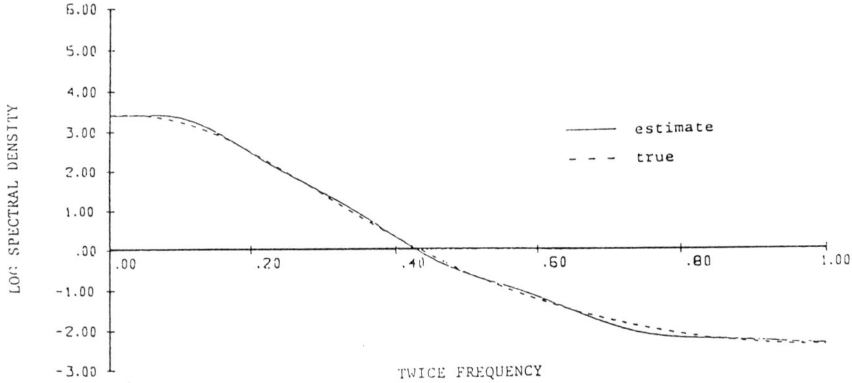

Figure 6. 1,024 points, first prior

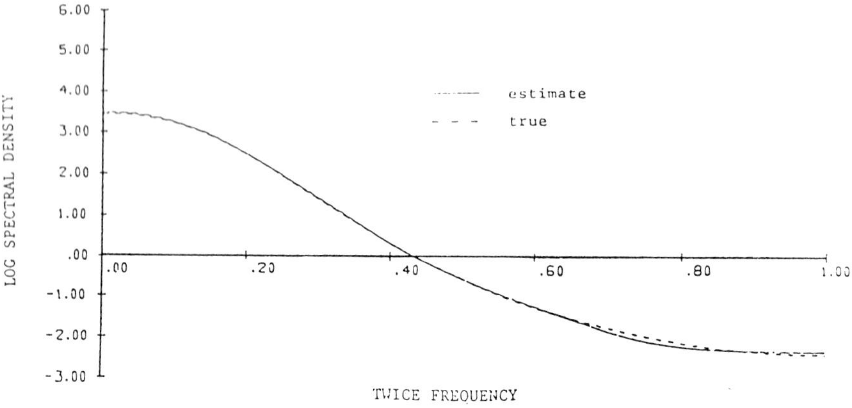

Figure 7. 1,024 points, fourth prior

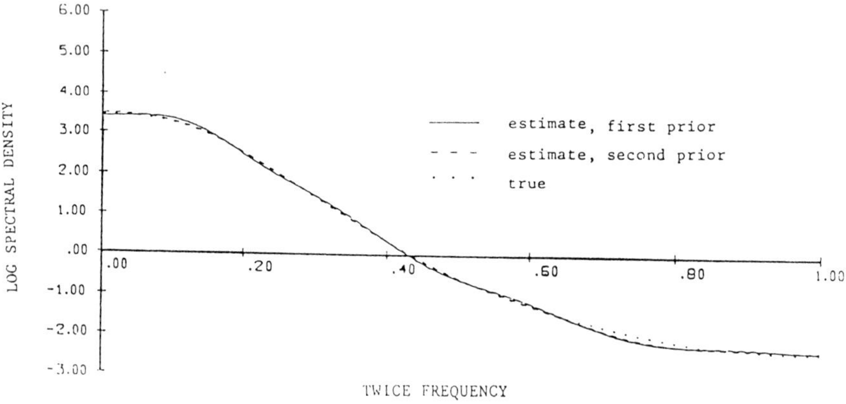

Figure 8. 1,024 points, first and fourth priors.

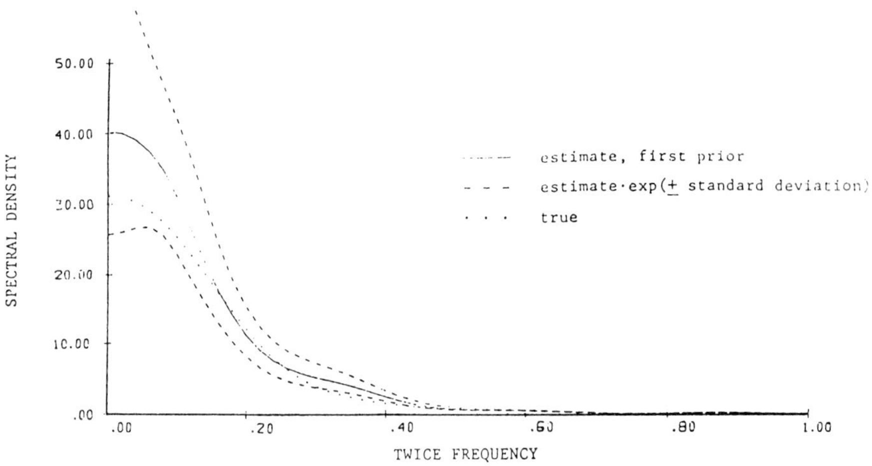

Figure 9. 256 points, first prior

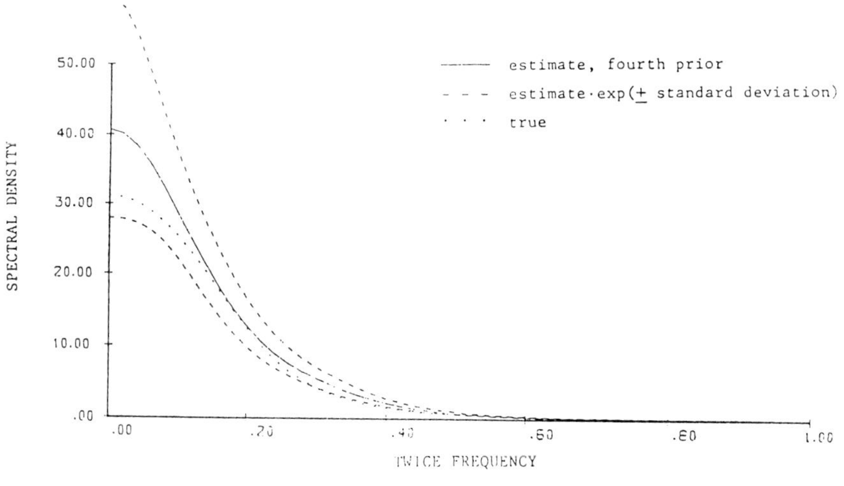

Figure 10. 256 points, fourth prior

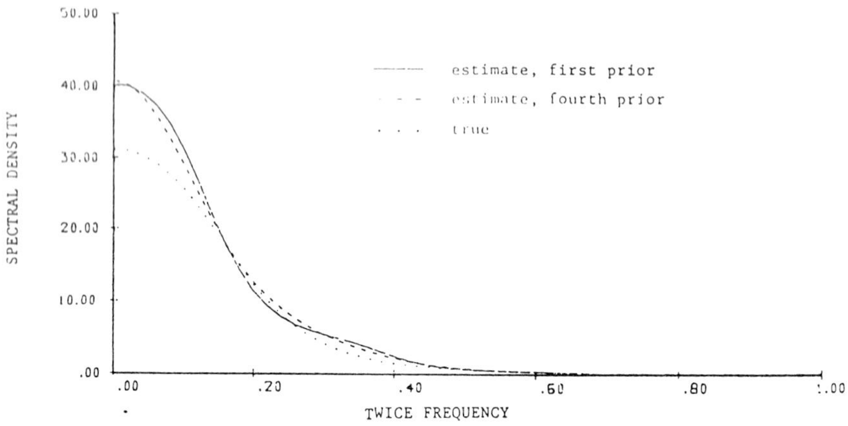

Figure 11. 256 points, first and fourth priors

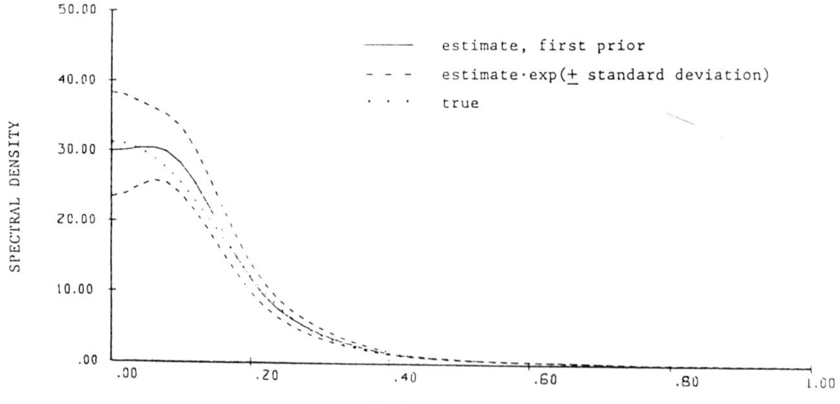

Figure 12. 1,024 points, first prior

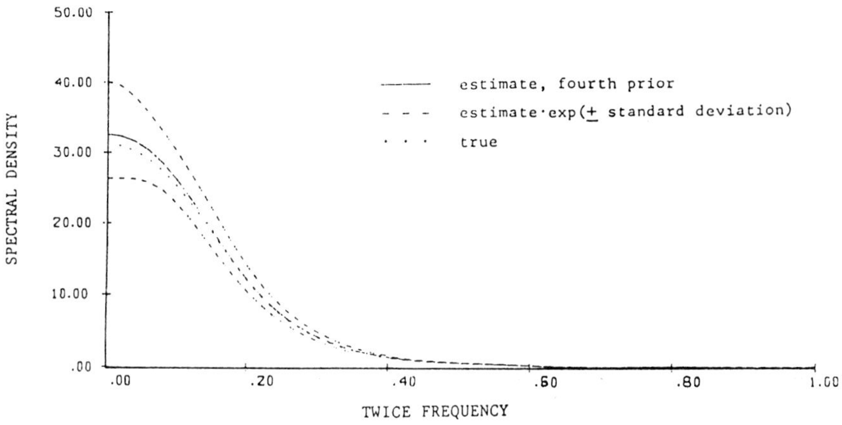

Figure 13. 1,024 points, fourth prior

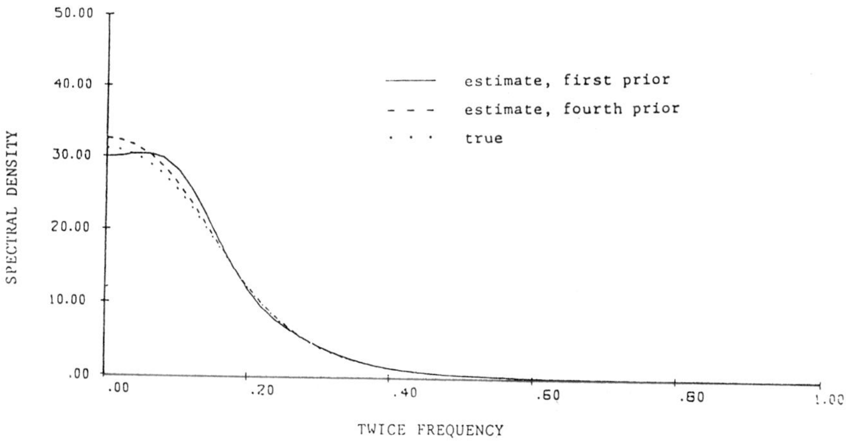

Figure 14. 1,024 points, first and fourth priors

PART III:

DETECTION AND SIGNAL EXTRACTION

SIGNAL DETECTION FOR SPHERICALLY EXCHANGEABLE (SE) STOCHASTIC PROCESSES

C.B. Bell
San Diego State University
San Diego, CA 92182

I. INTRODUCTION AND SUMMARY

Consider signal detection in the following context. The background pure noise (PN) is produced by nature's choosing a parameter value and then generating a time series according to a (stochastic process) law with that parameter value. This process may be repeated N times. If there is a signal present, what is generated is a noise-plus-signal $(N + S)$, time series. The statistical properties of the latter-type of time series are different from those of the former.

In the sequel, one will be concerned primarily with the case in which the parameter chosen is a variance and the time series subsequently generated is an i.i.d. zero-mean Gaussian time series with that variance. The variances are chosen by nature according to a positive cdf, corresponding to which will be a radial distribution.

The signal detection procedures developed below will be valid not only for the (conditional) Gaussian model, $\Omega(S-S-E)$, described above, but also for a wider family $\Omega(K-S-E)$ of pure noise (PN) stochastic processes. This is because for all $S-E$ (spherically exchangeable) distributions, the set of observed radii $\{R_j\}$ are sufficient, and the set of "angles" are independent of the radii and have identical uniform distributions. [See Section 2.C.]

Five different signal detection models will be explicitly treated. The treatment of these models will indicate how a variety of related signal detection problems can be treated.

The paper is divided into ten sections. In Section 2, the basic $S-E$ data models are given. Section 3 presents the five detection problems to be considered. Statistical preliminaries are presented for historical data and for cross-sectional data in Section 4. Detection procedures for the five problems are developed in Section 5 through 9.

Open problems and concluding remarks are given in Section 10.

II. DATA STRUCTURE OF THE BACKGROUND NOISE

It is assumed throughout that decisions are made on the basis of a finite number of observations. The actual time series being observed may be infinite, but one only uses a finite initial segment in the case of historical data; and a finite number of finite initial segments for cross-sectional data. In each case, the law, L, of the process determines the distribution of the initial segment, and vice versa.

The data to be considered is of the form $\underset{\sim 1}{Z}, \underset{\sim 2}{Z}, \ldots, \underset{\sim N}{Z}$, where $\underset{\sim r}{Z} = (Z_{r1}, \ldots, Z_{rk})$, and the probabilistic structures of interest are given below.

This work was supported by the Office of Naval Research through Grant No. N00014-8-C-0208.

(2A) Normal Spherical Exchangeability $\Omega(N-S-E)$

$\underset{\sim 1}{Z}, \ldots, \underset{\sim N}{Z}$ are i.i.d. $N(\underset{\sim}{0}, \sigma^2 \tilde{I}_k)$, i.e., the k components of each $\underset{\sim}{Z}$-vector are i.i.d. $N(0, \sigma^2)$. This is the original spherical structure of interest. The family of laws of all such time series is $\Omega(N-S-E)$.

(2B) Schoenberg (1938) Spherical Exchangeability $\Omega(S-S-E)$

$W_1, \ldots, W_N$ are i.i.d. $H(\cdot)$, with $H(0) = 0$, and, conditionally given $W_r = w_r$, $X_{r1}, \ldots, X_{rk}$ are i.i.d. $N(0, 1/w_r)$.

Definition 2.1. $H(\cdot)$ is called the mixing measure.

Here, one can say that each $\underset{\sim r}{Z}$ has a multivariate distribution which is a variance-mixture of zero-mean normal random samples. If $H(\cdot)$ is a one-point (i.e., degenerate) cdf then one has case (2A) above. The family of laws of all such time series is $\Omega(S-S-E)$.

(2C) Kelker Spherical Exchangeability, $\Omega(K-S-E)$

$\underset{\sim 1}{Z}, \underset{\sim 2}{Z}, \ldots, \underset{\sim N}{Z}$ are i.i.d. $\tilde{F}(\cdot)$, where $\underset{\sim j}{Z} = R_j \cdot \underset{\sim j}{V}$ with (1) R_j and $\underset{\sim j}{V}$ being independent; (2) $R_j \sim J(\cdot)$, where $J(0) = 0$; and (3) $\underset{\sim j}{V} = (V_{j1}, \ldots, V_{jk})$ being uniformly distributed over the hypersphere $S_1^* = \{\underset{\sim}{v}: \sum_1^k v_j^2 = 1\}$, with $R_j^2 = \sum_{s=1}^k Z_{js}^2$ and $V_{js} = \dfrac{Z_{js}}{R_j}$.

Definition 2.2. $J(\cdot)$ is called the *radial cdf.*
It is immediate that

Theorem 2.1.} $\Omega(N-S-E) \subset \Omega(S-S-E) \subset \Omega(K-S-E)$.
Schoenberg (1938) essentially proved that

Theorem 2.2. For infinite time series,

$$\Omega(S-S-E) = \Omega(K-S-E).$$

From Lord (1954); Kelker (1970); Bell, Avadhani and Woodroofe (1970); and Bell (1975, 1978), one has

Theorem 2.3. Let $\underset{\sim}{Z} = (Z_1, \ldots, Z_k)$ have a law L in $\Omega(K-S-E)$. Then for $1 \le m \le k-1$, $[\sum_1^m Z_j^2]\,[\sum_{m+1}^k Z_j^2]^{-1}(\frac{k-m}{m}) \sim F(m, k-m)$, Fisher's F-distribution, with m and $(k-m)$ degrees of freedom.

The salient point here is that there are non-Gaussian laws for which the F-distribution is valid.

These types of PN structures can be best illustrated by the following examples.

Example 2.1. $\Omega(N-S-E)$. Let $Z_{r1}, \ldots, Z_{r9}$ be i.i.d. $N(0{,}5.4)$; $\underset{\sim r}{Z} = (Z_{r1}, \ldots, Z_{r9})$; and $\underset{\sim 1}{Z}, \ldots, \underset{\sim 50}{Z}$ be i.i.d. The $\underset{\sim r}{Z}$ here are initial segments of zero-mean i.i.d. Gaussian time series.

Example 2.2. $\Omega(S-S-E)$. Let $W_1, \ldots, W_{20}$ be i.i.d. Exp (2.5). Further, given $W_r = w_r$, let $Z_{r1}, \ldots, Z_{r8}$ be conditionally i.i.d. $N(0, 1/w_r)$. Then, the unconditional density of $\underset{\sim r}{Z}$ is

$$\tilde{f}(z_1, \ldots, z_8) = \int_0^\infty (\frac{w}{2\pi})^4 \exp\{-\frac{w}{2}\sum_1^8 z_j^2\} \ (2.5) \exp\{-2.5w\}dw$$

$$= (120)\pi^{-4}[5.0 + \sum_1^8 z_j^2]^{-5}; \text{ and } \underset{\sim 1}{Z}, \ldots, \underset{\sim 20}{Z} \text{ are}$$

i.i.d. $\tilde{F}(\cdot, \ldots, \cdot)$.

In this example, each $R_j^2 = \sum_1^8 Z_{j\bullet}^2$, and the radial distribution $J(\cdot)$ satisfies

$$J(y) = P\{R_j \leq y\} = \int_0^\infty F^*(yw^2) \ (2.5) \exp\{-2.5w\}dw,$$

where $F^*(\cdot)$ is the X_8^2-cdf.

[Note: If $H(\cdot)$ were such that $P_H(\{5.2\}) = 1$, then $\tilde{f}(z)$ would be $(10.4\pi)^{-4} \exp\{-\frac{1}{10.4}\sum_1^8 z_j^2\}$ and the law would be in $\Omega(N-S-E)$.]

Example 2.3. $\Omega(K-S-E)$.

Let $\underset{\sim r}{Z} = (Z_{r1}, \ldots, Z_{r5})$ and $\underset{\sim 1}{Z}, \ldots, \underset{\sim 15}{Z}$ be i.i.d. $F_1(\cdot, \ldots, \cdot)$, with density of the form

$$\tilde{f}_1(z_1, \ldots, z_5) = \begin{cases} k_1^* \ \theta^{-1}(\sum_1^5 z_j^2)^{-2} \ , & 0 < \sum_1^5 z_j^2 < \theta^2 \\ 0 \ , & \text{otherwise} \end{cases}$$

Then, the radial distribution, $J_1(\cdot)$ is $U(0, \theta)$, i.e., $J_1'(y) = \theta^{-1}, 0 < y < \theta$.

If the common density for the $\underset{\sim}{Z}$'s is of the form

$$\tilde{f}_2(z_1, \ldots, z_5) = k_2^*\left[\sum_1^5 z_j^2\right]^{-2} \exp\{-\lambda\sqrt{\sum_1^5 z_j^2}\},$$

then the radial distribution satisfies

$$J_2(y) = 1 - \exp\{-\lambda y\}, \ y > 0.$$

Statistical detection procedures for these types of pure-noise data involve the observed radii, $R_1, \ldots, R_N$, which function as sufficient statistics; and the "angles" $\underset{\sim 1}{V}, \ldots, \underset{\sim N}{V}$ which are independent of the radii. [The $\underset{\sim}{V}$'s are not direction-angle vectors as in polar coordinates.]

For detection problems involving specific radial distributions one uses the $\{\underset{\sim}{R}_j\}$, while for detection problems involving the underlying spherical structure, the $\{\underset{\sim}{V}_j\}$ are employed.

The signal detection models to be treated in the sequel are given in the next section.

III. SOME SIGNAL DETECTION PROBLEMS

In each of the cases below, the *PN* (i.e., pure-noise) distribution or law, L, is described. Data reflecting a different stochastic process law would indicate that a signal is being received (in addition to the pure noise).

3.1. Detection Problems with Historical Data

For these problems, one assumes that the total data available is $\underset{\sim}{Z} = \{Z_r\colon 1 \leq 1 \leq k\}$. (In terms of the preceding section, one has $N = 1$.)

The problems are as follows:

$$PN_1\colon L(\{\underset{\sim}{Z}\}) \in \Omega(K-S-E)$$

This means the background pure noise is itself $S-E$.

$$PN_2\colon L(\{\underset{\sim}{Z} - a\}) \in \Omega(K-S-E) \text{ for some (unknown) a.}$$

For this case, one hypothesizes that the pure noise is $S-E$ about some unknown point a, possibly not zero. If "a" is known, subtraction leads to PN_1 above.

Certain other detection problems can only be handled when the data is cross-sectional, i.e., the data is of the form below.

$$\tilde{Z} = \begin{pmatrix} \underset{\sim}{Z}_1 \\ \cdot \\ \cdot \\ \cdot \\ \underset{\sim}{Z}_N \end{pmatrix} \text{ where } \underset{\sim}{Z}_r = (Z_{r1}, \ldots, Z_{rk})$$

3.2. Detection Problems with Cross-Sectional Data

$$PN_3\colon L(\underset{\sim}{Z}) = L_0 \in \Omega(K-S-E)$$

Here, one gives a specific pure-noise law. Any other stochastic behavior of the data indicates signal is present. $\underset{\sim}{Z}_1, \ldots, \underset{\sim}{Z}_n$ are i.i.d. and the associated i.i.d. radii $R_1, \ldots, R_n$ are used in the decision procedures.

$$PN_4\colon L(\{\underset{\sim}{Z} - a\}) = L_0 \in \Omega(K-S-E) \text{ for some a.}$$

This situation refers to a family of PN-laws indexed by "a", a nuisance parameter. The decision procedures will involve radii modified by "estimates" of "a".

The final detection model to be considered is a two-sample model. The observer observes two independent time series $\{X_r\}$ and $\{Y_r\}$, and the pure-noise situation entails their stochastic laws being equal. [See Bell (1964), Model II.]

The data here is

$$\underset{\sim}{\tilde{Z}} = \begin{pmatrix} \underset{\sim 1}{Z} \\ \cdot \\ \cdot \\ \cdot \\ \underset{\sim N}{Z} \end{pmatrix}$$

where

$$\underset{\sim r}{Z} = \begin{cases} (X_{r1}, \ldots, X_{rk}), & \text{for } 1 \leq r \leq m \\ (Y_{r1}, \ldots, Y_{rk}), & \text{for } m+1 \leq r \leq N \end{cases}$$

$$PN_5\text{:} \ L(\{X_r\}) = L(\{Y_r\})$$

The mechanisms for handling these five detection problems (and some related problems) are given in the section below.

IV. STATISTICAL PRELIMINARIES: SUFFICIENT STATISTICS AND STATISTICAL NOISE

For developing the methodology it is easier to consider first $\Omega(S-S-E)$, since the family consists of laws involving conditional Gaussian distributions. Nuisance parameters and $M-S-S$'s (minimal sufficient statistics) can be handled in traditional ways. Then since the "angles" $\{\underset{\sim j}{V}\}$ have the same behavior for $\Omega(N-S-E)$, $\Omega(S-S-E)$ and $\Omega(K-S-E)$, the decision procedures developed can be extended from $\Omega(S-S-E)$ to $\Omega(K-S-E)$.

One recalls that the above-mentioned extension is not necessary if the data vectors are initial segments of infinite $S-E$ time series. For such cases (see Theorem 2.2) $\Omega(S-S-E) = \Omega(K-S-E)$.

To these ends, one now develops the mechanisms with $\Omega(S-S-E)$ in mind.

(4A) Basic Statistics

Let $\tilde{Z}$ be a data matrix with law L' in a family Ω' admitting a $M-S-S$ (minimal sufficient statistic) $S(\tilde{Z})$.

Definition 4.1. Let $\delta(\tilde{Z}) = [S(\tilde{Z}), N(\tilde{Z})]$ where (1) $\delta(\cdot)$ is 1-1 a.e.; (2) $S(\tilde{Z})$ and $N(\tilde{Z})$ are independent. Then (a) $\delta(\cdot)$ is called the BDT (basic data transformation) for Ω'; and (b) $N(\tilde{Z})$ is called the $M-S-N$ (maximal statistical noise) for Ω'.

[One should note that for any given family Ω', $S(\tilde{Z})$, $N(\tilde{Z})$ and $\delta(\cdot)$ need not be unique.]

These entities are used in contructing families of detection statistics with certain desirable properties.

(4B) Distribution-Free Statistics

Definition 4.2.

(1) A statistic $T(\tilde{Z})$ is NPDF wrt Ω', if there exists a cdf $Q(\cdot)$ such that $P\{T(\tilde{Z}) \leq y \mid L\} = Q(y)$ for all y and for all L in Ω'.

(2) A *family* of statistics $\{T^*(\tilde{Z}, L)\colon L \in \Omega'\}$ is PDF wrt Ω', if there exists a cdf $Q^*(\cdot)$ such that

$$P\{T^*(\tilde{Z}, L) \leq y \mid L\} = Q^*(y), \text{ for all } y \text{ and for all } L \text{ in } \Omega'.$$

RULE OF THUMB:

4.A For detection problems in which the PN-distribution involves a specific law, L_0, employ statistics of the form $T^*(\tilde{Z}, L_0) = \psi^*[S(\tilde{Z}), L_0]$. This entails using the data only via the $M-S-S$. Such statistics are PDF wrt Ω'.

4.B For detection problems which are concerned with the general structure of the PN-family, employ statistics of the form $T(\tilde{Z}) = \psi[N(\tilde{Z})]$. This entails using the data only through the $M-S-N$. Such statistics are NPDF wrt Ω'.

Example 4.1 Let $L_H \in \Omega(S-S-E)$ and $\underset{\sim}{Z} = (Z_1, \ldots, Z_7)$. This means nature has chosen $W = w$, where $W \sim H(\cdot)$, and given $W = w$, $Z_1, \ldots, Z_7$ are conditionally i.i.d. $N(0, \frac{1}{w})$. $R = (\sum_1^7 z_j^2)^{1/2}$ is the M-S-S for $H(\cdot)$, which determines and is determined by L (See Bell, et al, 1970).

$$\underset{\sim}{V} = (\frac{Z_1}{R}, \ldots, \frac{Z_7}{R}) \text{ is M–S–N; and}$$

$$S(\underset{\sim}{Z}) = (R, \underset{\sim}{V}), \text{ where } S(\underset{\sim}{Z}) = R \text{ and } N(\underset{\sim}{Z}) = \underset{\sim}{V}. \text{ Now, let}$$

$$T_1(\underset{\sim}{Z}) = V_1 = \frac{Z_1}{R} \text{ and } T_2(\underset{\sim}{Z}, L_H) = J_L(R),$$

where $J_L(y) = P\{R \leq y \mid H\} = P\{R \leq y \mid L_h\}$, i.e., J_L is the radial cdf. Then $T_1(\cdot)$ is NPDF wrt $\Omega(S-S-E)$ with cdf $Q(z) = 1 - F(6,1;\frac{(1-y^2)}{6-y^2})$ for $y > 0$; and $T_2(\cdot\,,\cdot)$ is PDF wrt $\Omega(S-S-E)$. (See Ex. 4.3.)

Now for $\Omega(K-S-E)$, R is the M-S-S for the radial distribution, and $\underset{\sim}{V}$ is M-S-N. Therefore, T_1 is NPDF wrt $\Omega(K-S-E)$, and T_2 is PDF wrt $\Omega(K-S-E)$.

[NOTE: For analyses somewhat different from that of the sequel, one needs a polar coordinate model, angular distributions, and the relations between joint densities, radial distributions and characteristic functions. These are given in Lord (1954); Kelker (1969);

Smith (1971); Bell and Smith (1970); and Ahmad (1975).]

(4C) Extraneous Statistical Noise (E-S-N)}

One further statistical tool involves the use of E-S-N. The essentials of this method have been employed by several authors, e.g., Durbin (1961); and Bell and Doksum (1965).

Definition 4.3}. Let $\underset{\sim}{Z} = (Z_1, \ldots, Z_k)$ be data governed by law L in Ω' with M-S-S, $S(\underset{\sim}{Z})$; M-S-N, $N(\underset{\sim}{Z})$; and BDT, $\delta(\cdot)$. Let $\underset{\sim}{Y} = (Y_1, \ldots, Y_k)$ be independent of $\underset{\sim}{Z}$ and be governed by L_1 in Ω'. Define $\underset{\sim}{Y'}$ to be $[Y_1', \ldots, Y_k'] = \delta^{-1}[S(\underset{\sim}{Y}), N(\underset{\sim}{Z})]$.

(1) $\underset{\sim}{Y}$ is called E-S-N (extraneous statistical noise); and

(2) $\underset{\sim}{Y'}$ is called R-S-N (randomized statistical noise).

Paralleling the proofs of the aforementioned authors, one can prove

Theorem 4.1 (Randomized Noise Theorem). $\underset{\sim}{Y'} \overset{d}{=} \underset{\sim}{Y}$. [See Bell, 1984.]

This result illustrates a method of imposing a known (usually tractable), distribution on a problem in which the cdf of the data is unknown. It is particularly useful when the distribution of the M-S-N, $N(\underset{\sim}{Z})$, is relatively intractable.

Example 4.2. Let $\underset{\sim}{Z} = (Z_1, \ldots, Z_{20})$ be i.i.d. $N(0,u^*)$ where $u^* > 0$ is unknown. Let $\underset{\sim}{Y} = (Y_1, \ldots, Y_{20})$ be independent of $\underset{\sim}{Z}$ and i.i.d. $N(0,1)$. Then the BDT is such that $\delta(\underset{\sim}{Z}) = [S(\underset{\sim}{Z}), N(\underset{\sim}{Z})]$, where $S(\underset{\sim}{Z}) = R = \sqrt{\sum_1^{20} Z_j^2}$, and $N(\underset{\sim}{Z}) = \underset{\sim}{V}$, with $V_j = \frac{Z_j}{R}$. $\underset{\sim}{V}$ is uniformly distributed over the hypersphere $S^*(1)$ [See Section 2C]. $\delta^{-1}[S(\underset{\sim}{Y}), N(\underset{\sim}{Z})] = \frac{R^*}{R}(Z_1, \ldots, Z_{20}) = (Y_1', \ldots, Y_{20}') = \underset{\sim}{Y'}$, where $(R^*)^2 = \sum_1^{20} Y_j^2$. Then, $Y_1', \ldots, Y_{20}'$ are i.i.d. $N(0,1)$.

(4D) Nuisance Parameters and Kolmogorov-Smirnov Statistics

The next set of statistical tools of this section involve modified K-S (Kolmogorov-Smirov) statistics. The original statistic of this class was (Kolmogorov (1933)).

$$D(F_0; n) = sup\ |F_n(z) - F_0(z)|, \text{ where}$$

$$F_n(z) = \frac{1}{n} \sum_1^n \in(z - Z_j), \text{ where } Z_1, \ldots, Z_n \text{ are i.i.d.}$$

$F_0(\cdot)$, continuous; and $\in(u) = 1$, if $u \geq 0$, 0, if $u < 0$.

For several of the problems of the sequel, $F_0(\cdot)$, is not known completely. It is known to be a member of a family of cdfs; and hence, there is a nuisance parameter involved. That is, $F_0(\cdot) \in \Omega' = \{F(\cdot;\theta): \theta \in \Theta\}$.

For such situations, one uses extensions of the ideas of Lilliefors (1967, 1969), Srinivasan (1970), Choi (1980), Bell (1984) and Bell and Mason (1985).

Definition 4.3. Let $\underset{\sim}{Z} = (Z_1, \ldots, Z_n)$ be i.i.d. $F_0 \in \Omega' = \{F(\cdot;\theta): \theta \in \Theta\}$, where Ω' admits a M-S-S, $S(\underset{\sim}{Z})$, and a MLE, $\hat{\theta}$, of θ. Let $\hat{F}_n(y) = F(y;\hat{\theta})$ for all y; and $\tilde{F}_n(y) = E(F_n(y) \mid S(\underset{\sim}{Z}))$, for all y.

(1) $\hat{D}_n = \sup_y |F_n(y) - \hat{F}_n(y)|$, is the Lilliefors-type statistic for Ω'; and

(2) $\tilde{D}_n = \sup_y |F_n(y) - \tilde{F}_n(y)|$ is the Srinivasan-type statistic for Ω'.

Example 4.3. Let $\underset{\sim}{Z} = (Z_1, \ldots, Z_{15})$ be i.i.d. $N(0,\theta)$. Then, the M-S-S, $S(\underset{\sim}{Z}) = R$, and the M-S-N, $N(\underset{\sim}{Z}) = \underset{\sim}{V}$, where $R^2 = \sum_1^{15} Z_j^2$, $V_j = \frac{Z_j}{R}$, and $\underset{\sim}{V} = (V_1, \ldots, V_{15})$. Then, $\hat{F}_{15}(y) = \Phi(\frac{y\sqrt{15}}{R})$, since $\hat{\theta} = \frac{1}{15}R^2$. Further $\tilde{F}_{15}(y) = E\{F_n(y) \mid R\} = P\{Z_1 \le y \mid R\} = H(14; R, x)$ for $-R < y < R$, where

$$H(m;R,x) = \begin{cases} \frac{1}{2} F(m;1;\frac{R^2 - x^2}{(m-1)x^2}) & , \quad -R \le x < 0 \\ 1 - \frac{1}{2} F(m;1;\frac{R^2 - x^2}{(m-1)x^2}) & , \quad 0 \le x \le R \end{cases}$$

The Lilliefors-type statistic is, then,

$\hat{D}_{15} = \sup_y |F_{15}(y) - \Phi(\frac{y\sqrt{15}}{R})|$, and the Srinivasan–type statistic is

$\tilde{D}_{15} = \sup_{-R < y < R} |F_{15}(y) - \tilde{F}_{15}(y)|$.

Example 4.4. Let $\underset{\sim}{Z} = (Z_1, \ldots, Z_9)$ be i.i.d. $N(\underset{\sim}{\theta})$, where $\underset{\sim}{\theta} = (\mu, \sigma^2)$. Then $S(\underset{\sim}{Z}) = (\bar{Z}, S_Z)$, where $S_Z^2 = \frac{1}{8}\sum_1^9 (Z_i - \bar{Z})^2$; and $N(\underset{\sim}{Z}) = (\frac{Z_1 - \bar{Z}}{S_Z}, \ldots, \frac{Z_9 - \bar{Z}}{S_Z}) = (W_1, \ldots, W_9)$. Here $\hat{D}_9 = \sup_y |F_9(y) - \Phi(\frac{y - \bar{Z}}{S_Z})|$. This is the statistic studied by Lilliefors (1967).

$$\tilde{D}_9 = \sup |F_9(y) - \tilde{F}_9(y)|, \text{ where } \tilde{F}_9(\cdot),$$

is given by Srinivisan (1970).

What is of importance here is that each of the statistics $\hat{D}_{15}$, $\tilde{D}_{15}$, $\hat{D}_9$ and $\tilde{D}_9$ (of the preceeding two examples) is a function of the data only through the relevant M-S-N. In fact, one can easily derive

(a) $\hat{D}_{15} = \sup_x |\frac{1}{15}\sum_1^{15} \in(x - V_j) - \Phi(x\sqrt{15})|$

(b) $\tilde{D}_{15} = \sup_{-1 < x < 1} |\frac{1}{15}\sum_1^{15} \in(x - V_j) - H(14; 1; 1, x)|$

and

(c) $\hat{D}_9 = \sup_x \left| \frac{1}{9} \sum_1^9 \in(x - V_j) - \Phi(x) \right|$

The final statistical tool of this section involves the use of Helmert matrices.

(4E) Helmert Matrices, M-S-S's and M-S-N

Definition 4.4. An $(N \times N)$ square matrix $\tilde{H}_N = \{h_{ij}\}$ is called a Helmert matrix of order N if

$$h_{ij} = \begin{cases} \frac{1}{\sqrt{N}} & \text{for} \quad i = N \\ \frac{1}{\sqrt{i(i-1)}} & \text{for} \quad 1 \leq i \leq N-1,\ 1 < j + \leq i - 1 \\ \frac{-(i-1)}{\sqrt{i(i-1)}} & \text{for} \quad 1 < i \leq N-1,\ j = i \\ 0 & \text{for} \quad 1 \leq i \leq N-1,\ i+1 \leq j \leq N \end{cases}$$

It is clear that each $\tilde{H}_N$ is orthogonal, and that

Theorem 4.1. If $\underset{\sim}{Z} = (Z_1, \ldots, Z_N)$ is governed by a law L in $\Omega(K{-}S{-}E)$ and $\underset{\sim}{X} = \underset{\sim}{Z}\, \tilde{H}_N^T$, then $\underset{\sim}{X} \overset{d}{=} \underset{\sim}{Z}$.

The Helmert matrix can in some circumstances be used to construct the M-S-S and M-S-N.

Example 4.5. (Gaussian Random Sample). Let $\underset{\sim}{Z} = (Z_1, \ldots, Z_5)$ be i.i.d. $N(\mu, \sigma^2)$ and let $\underset{\sim}{Y} = (Y_1, \ldots, Y_5) = \underset{\sim}{Z}\, \tilde{H}_5^T$, where

$$\tilde{H}_5 = \begin{pmatrix} \frac{1}{\sqrt{2}} & -\frac{1}{\sqrt{2}} & 0 & 0 & 0 \\ \frac{1}{\sqrt{6}} & \frac{1}{\sqrt{6}} & \frac{-2}{\sqrt{6}} & 0 & 0 \\ \frac{1}{\sqrt{12}} & \frac{1}{\sqrt{12}} & \frac{1}{\sqrt{12}} & \frac{-3}{\sqrt{12}} & 0 \\ \frac{1}{\sqrt{20}} & \frac{1}{\sqrt{20}} & \frac{1}{\sqrt{20}} & \frac{1}{\sqrt{20}} & \frac{-4}{\sqrt{20}} \\ \frac{1}{\sqrt{5}} & \frac{1}{\sqrt{5}} & \frac{1}{\sqrt{5}} & \frac{1}{\sqrt{5}} & \frac{1}{\sqrt{5}} \end{pmatrix}$$

Then, Y_1, Y_2, Y_3, Y_4 and Y_5 are independent, with $Y_j \sim N(0, \sigma^2)$ for $1 \leq j \leq 4$, and $Y_5 \sim N(\mu\sqrt{5}, \sigma^2)$. One has, then, that the M-S-S for (μ, σ^2) is $R(\underset{\sim}{Z}) = (Y_5, R^*)$, where $(R^*)^2 = \sum_1^4 Y_j^2$ and the M-S-N is $N(Z) = (V_1, \ldots, V_4)$ where $V_j = \frac{Y_j}{R^*}$.

One can now treat the first detection problem.

V. S-E BACKGROUND NOISE

$$PN_1:\ L \in \Omega(K-S-E) \text{ vs } N + S_1:\ L \notin \Omega(K-S-E)$$

As previously mentioned, the technique will be to develop the methodology for $\Omega(S-S-E)$, and extend its validity to $\Omega(K-S-E)$.

For $\Omega(S-S-E)$, one historical "look" yields $\underset{\sim}{Z} = (Z_1, \ldots, Z_N)$ which are i.i.d. $N(0, (w^*)^{-1})$, given $W = w^*$, where $W \sim H(\cdot)$, with $H(0) = 0$.

Then, one has

Theorem 5.1. Conditionally given $W = w^*$,

(a) the M-S-S is $S(\underset{\sim}{Z}) = R$;

(b) the M-S-N is $N(\underset{\sim}{Z}) = \underset{\sim}{V} = (\frac{Z_1}{R}, \ldots, \frac{Z_N}{R})$;

(c) the BDT is $\delta(\underset{\sim}{Z}) = (R, \underset{\sim}{V})$; and

(d) The MLE of w^* is $[\frac{1}{N} R^2]^{-1}$.

From these entities, one derives the decision rule based on the relevant statistics of Section 4.

Decision Rule 5.1: Decide $N + S_1$ iff

$$\hat{D}_N = \sup_y |F_N(y) - \Phi(\frac{y\sqrt{N}}{R})| > \hat{d}(\alpha, N).$$

Decision Rule 5.2: Decide $N + S_1$ iff

$$\tilde{D}_N = \sup_y |F_N(y) - H(N-1;\ R, y)| > \tilde{d}(\alpha, N).$$

Decision Rule 5.3: Decide $N + S_1$ iff

$$T = [\sum_1^m Z_j^2]\,[\sum_{m+1}^N Z_j^2]^{-1}\ (\frac{N-m}{m}) < f' \text{ or } > f'.$$

[The $f'-$ and $f'-$ values are found in a Fisher's F-table for m and $(N-m)$ degrees of freedom].

Decision Rule 5.4: Decide $N + S_1$ iff

$$D_N' = \sup_z |\frac{1}{N}\sum_1^N \in (z - Y_j') - \Phi(z)| > d(\alpha, N), \text{ where } \underset{\sim}{Y} = (Y_1, \ldots, Y_N)$$

is E-S-N and $\underset{\sim}{Y'} = (Y_1, \ldots, Y_N)$ is R-S-N as in Section 4C. The $d(\alpha, N)-$ value are found in a standard one-sample K-S table. [Note: If the data available is cross-sectional data, one might make the adaptations in decision rules suggested at the end of Section 6.]

It should be noted that each of the decision rules of this section involves the data solely via the M-S-N, $\underset{\sim}{V}$, and hence is NPDF. Also $\underset{\sim}{V}$ is the M-S-N for $\Omega(S-S-E)$ as well as for $\Omega(K-S-E)$. Hence, one has

Theorem 5.2. (1) Each of the statistics $\hat{D}_N$, $\tilde{D}_N$, T, D_N' can be written in the form $\psi[\underset{\sim}{V}]$, i.e., $\hat{D}_N = \psi_1[\underset{\sim}{V}]$; $\tilde{D}_N = \psi_2[\underset{\sim}{V}]$, $T = \psi_3[\underset{\sim}{V}]$ and $D_N' = \psi_4[\underset{\sim}{V}]$.

(2) Hence, each of these statistics is NPDF wrt $\Omega(K-S-E)$.

The next detection problem involves two nuisance parameters from the point of view of $\Omega(S-S-E)$.

VI. BACKGROUND NOISE S-E ABOUT AN UNKNOWN POINT, 'a'

$\underline{PN_2}$: $L(\{Z - a\}) \in \Omega(K-S-E)$ for some point 'a' vs

$\underline{N + S_2}$: $L(\{Z - a\}) \notin \Omega(K-S-E)$ for any point 'a'.

From the $\Omega(S-S-E)$ point of view, the historical data is $Z = (X_1, \ldots, X_N)$, which are c.i.i.d. $N(a, (w^*)^{-1})$, given $W = w^*$.

Theorem 6.1. The basic statistics (conditionally) are then

(a) M-S-S: $S(\underset{\sim}{Z}) = (\bar{X}, S_x)$;

(b) M-S-N: $N(\underset{\sim}{Z}) = (U_1, \ldots, U_N) = \underset{\sim}{U}$, where $U_j = \dfrac{X_j - \bar{X}}{S_X}$;

(c) BDT: $\delta(\underset{\sim}{Z}) = (\bar{X}, S_X, \underset{\sim}{U})$; and

(d) MLE of (a, w^*): $(\bar{X}, N[(N-1)S_X^2]^{-1})$.

Some important decision rules are then as follows:

Decision Rule 6.1. Decide $N + S_2$ iff

$$\overset{*}{D}_N = \sup_y |F_{N(y)} - \Phi(\frac{y - \bar{X}}{S_X})| > \overset{*}{d}(\alpha, N).$$

[The Lilliefors (1967) table yields critical values. See also Mason, 1984 and Mason and Bell, 1986.]

Decision Rule 6.2. Decide $N + S_2$ iff

$$\tilde{\tilde{D}}_N = \sup_y |F_N(y) - \tilde{F}_N(y)| > \tilde{\tilde{d}}(\alpha, N).$$

[Both $\tilde{F}_N(\cdot)$ and the critical values are given by Mason, 1984 and Mason and Bell, 1986.]

Decision Rule 6.3. Decide $N + S_2$ iff

$$D_N' = \sup_y \left| \frac{1}{N} \sum_1^N \in (y - Y_j') - \Phi(y) \right| > d(\alpha, N)$$

where $Y_j' = \overline{Y} + S_Y(\frac{X_j - \overline{X}}{S_X})$, and $Y_1, \ldots, Y_N$ is E-S-N.
[The $d(\alpha, N)$ values are from the standard one-sample K-S table.]

Some different procedures result from employing Helmert matrices here.

Theorem 6.2. Let $\underset{\sim}{Y} = (Y_1, \ldots, Y_N) = \underset{\sim}{Z} \tilde{H}_N^T$. Then, conditionally, given $W = w^*$, one has

(a) $Y_1, \ldots, Y_N$ are independent;
(b) $Y_j \sim N(0, 1/w^*)$ for $1 \leq j \leq N - 1$;
(c) $Y_N \sim N(a\sqrt{N}, \frac{1}{w^*})$;
(d) (Y_N, R^*) is the M-S-S; where $R^* = [\sum_1^{N-1} Y_j^2]^{-1/2}$;
(e) $\underset{\sim}{V} = (V_1, \ldots, V_{N-1})$ is the M-S-N, with $V_j = \frac{Y_j}{R^*}$; and
(f) $(\frac{Y_N}{\sqrt{N}}, \frac{N}{(R^*)^2})$ is the MLE of (a, w^*).

The new decision rules are based on an F-ratio and the empirical cdf $F_{N-1}^*(\cdot)$, where $F_{N-1}^*(z) = \frac{1}{N-1} \sum_{j=1}^{N-1} \in (z - Y_j)$.

Decision Rule 6.4. Decide $N + S_2$ iff

$$\hat{D}_{N-1} = \sup_z \left| F_{N-1}^*(z) - \Phi(\frac{z\sqrt{N-1}}{R^*}) \right| > \hat{d}(\alpha, N-1).$$

Decision Rule 6.5. Decide $N + S_2$ iff

$$\tilde{D}_{N-1} = \sup_z \left| F_{N-1}^*(z) - H(N-2;\ R^*, z) \right| > \tilde{d}(\alpha, N-1).$$

Decision Rule 6.6. Decide $N + S_2$ iff

$$D_{N-1} = \sup_z \left| \frac{n}{N-1} \sum_1^{N-1} \in (z - X_j') - \Phi(z) \right| > d(\alpha, N-1).$$

where $\underset{\sim}{X} = (X_1, \ldots, X_{N-1})$ is E-S-N and $X_j' = \frac{R^{**}}{R^*} X_j$ and $(R^{**})^2 = \sum_1^{N-1} X_j^2$. [See Decision Rule 5.4 for critical values.]

Decision Rule 6.7. Decide $N + S_2$ iff

$$T' = [\sum_1^m Y_j^2] [\sum_{m+1}^{N-1} Y_j^2]^{-1} (\frac{N - m - 1}{m}) > f' \text{ or } < f'$$

[See Decision Rule 5.3 for critical values.]

One should note that in the event that cross-sectional data is available, some adjustments should be made in the decision rules of this section and the preceeding section. One such adjustment is suggested by the following development, based on Bell, 1975.

Let $\underset{\sim 1}{Z}, \underset{\sim 2}{Z}, \ldots, \underset{\sim N}{Z}$ be i.i.d. initial k-segments of time series with common law L in $\Omega(K-S-E)$. Then one has

Theorem 6.8. (1) $P\{(\underset{\sim 1}{Z}, \underset{\sim 2}{Z}) = (Z_{11}, \ldots, Z_{1k}, Z_{21}, \ldots, Z_{2k})$ is $S-E\} = 0$; and (2) $(\bar{Z}_{.1}, \ldots, \bar{Z}_{.k})$ is $S-E$, where $\bar{Z}_{.j} = \frac{1}{N}\sum_{r=1}^{N} Z_{rj}$.

This result suggests that the decision rules of Section 5 and 6 can be applied to the mean vectors of cross-sectional data.

With so many decison rules being considered, it is natural to ask which rules are better. More specifically, one might ask how decision rules based on the Z's compare with those based on the Y's of this section. No definitive answers are known at this time. [See Section 10.]

One now turns to detection problems for which cross-sectional data is necessary and available.

VII. GOODNESS-OF-FIT DETECTION PROBLEMS

$\underline{PN_3}$: $L = L_0 \in \Omega(K-S-E)$ vs $\underline{N + S_3}$: $L \neq L_0$

It is convenient to write the data in matrix form.

$$\tilde{Z} = \begin{pmatrix} Z_{11} & \cdot & \cdot & \cdot & \cdot & Z_{1k} \\ \cdot & & & & & \\ \cdot & & & & & \\ \cdot & & & & & \\ Z_{N1} & \cdot & \cdot & \cdot & \cdot & Z_{Nk} \end{pmatrix}$$

As usual one views the problem first from the $\Omega(S-S-E)$ viewpoint. This means that $W_1, \ldots, W_N$ are i.i.d. $H(\cdot)$ with $H(0) = 0$; and $Z_{j1}, \ldots, Z_{jk}$ are c.i.i.d. $N(0, [w_j^*]^{-1})$, given $W_j = w_j^*$; for $j = 1,2,\ldots,N$.

One now defines $R_j^2 = \sum_{m=1}^{k} Z_{jm}^2$; $\underset{\sim j}{V} = (V_{j1}, \ldots, V_{jk})$ where $V_{jm} = \frac{Z_{jm}}{R_j}$; and $G_N(z) = \frac{1}{N}\sum_{1}^{N} \in (z - R_j)$.

It can be proved that

Theorem 7.1. Conditionally, given $W_j = w_j^*$ for $1 \leq j \leq N$,

(1) the M-S-S is $S(\tilde{Z}) = [R(1), \ldots, R(N)]$, the ordered radii;
(2) The M-S-N is $N(\tilde{Z}) = [R(R_1), \ldots, R(R_N), \underset{\sim 1}{V}, \ldots, \underset{\sim N}{V}]$; and
(3) $R_1, \ldots, R_N$ are i.i.d. $G_0(\cdot)$, where $G_0(z) = P\{R_1 \leq z \mid L_0\}$, i.e., $G_0(\cdot)$ is the radial

cdf.

A pertinent decision rule is as follows.

Decision Rule 7.1. Decide $N + S_3$ iff

$$D_N = \sup_z |G_N(z) - G_0(z)| > d(\alpha, N).$$

[See Decision Rules 5.4 and 6.6 for critical values.]

One aspect of the decision rule above worth knowing is the exact relationship between L_0, the law in $\Omega(S-S-E)$ and $H_0(\cdot)$, the mixing measure and cdf of the W's; and $\tilde{f}_0(\cdot, \ldots, \cdot)$ the joint density function of $\underset{\sim}{Z} = (Z_1, \ldots, Z_N)$. It can be proved that

Theorem 7.2. (1) $G_0(z) = \int_0^\infty N_k(wz^2)dH_0(z)$ where $N_k(\cdot)$ is the cpf of a X_k^2-distribution. (2) For fixed N, each of $G_0(\cdot)$, L_0, $H_0(\cdot)$ and $\tilde{f}_0(\cdot, \ldots, \cdot)$ is determined by any other. [See, e.g., Kelker (1971), Bell, et al (1970), Bell (1975).] This means that the detection problem $\underline{PN_3}$: $L = L_0$ is equivalent to $\underline{PN_3^*}$: $G = G_0$; and, hence, the decision rule is "completely relevant."

A related detection problem is treated in the next section.

VIII. GOODNESS-OF-FIT WITH A NUISANCE PARAMETER

PN_4: $L(\{Z_r - a\}) = L_0 \in \Omega(K-S-E)$ for some 'a'(possibly non-zero)

vs

$N + S_4$: For each a , $L(\{Z_r - a\}) \neq L_0$.

For the data matrix $\tilde{Z}$ as in Section 7, one forms $\tilde{Y} = \tilde{Z}\,\tilde{H}_N^T =$

$$\begin{pmatrix} Y_{11} & \cdot & \cdot & \cdot & Y_{1k} \\ \cdot & & & & \\ \cdot & & & & \\ \cdot & & & & \\ Y_{N1} & \cdot & \cdot & \cdot & Y_{Nk} \end{pmatrix}$$

For constructing the M-S-S and M-S-N one needs

(a) $Y_j^* = Y_{jk}$ for $1 \leq j \leq N$;

(b) $(R_j^*)^2 = \sum_{s=1}^{k-1} Y_{js}^2$;

(c) $\underset{\sim}{Y}^* = [Y^{(1)}, Y^{(2)}, \ldots, Y^{(N)}]$ where $Y^{(s)}$ is associated with $R^*(s)$;

(d) $\underset{\sim}{R}^* = [R^*(1),\ldots, R^*(N)]$, the ordered R^*'s;

(e) $\underset{\sim j}{V} = (V_{j1}, \ldots, V_{j,k-1})$, where $V_{js} = \dfrac{Y_{js}}{R_j^*}$.

One has then,

Theorem 8.1. Conditionally, given $W_j = w_j^*$ for $1 \le j \le N$,

(a) $S(\tilde{Z}) = S(\tilde{Y}) = [\underset{\sim}{Y}^*, \underset{\sim}{R}^*]$; and

(b) $N(\tilde{Z}) = N(\tilde{Y}) = [R(Y_1^*),..., R(Y_N^*); R(R_1^*),..., R(R_N^*); \underset{\sim 1}{V}, \ldots, \underset{\sim N}{V}]$

For the decision procedure, one needs the cdf of R_1^*. Call it $G_0^*(\cdot)$ where $G_0^*(z) = P\{R_1^* \le z\}$.

Theorem 8.2. (a) $G_0^*(\cdot)$ determines and is determined by L_0;

(b) $R_1^*, \ldots, R_N^*$ are i.i.d. G_0^*

The decision rule is then

Decision Rule 8.1. Decide $N + S_4$ iff

$$D_{N-1} = \sup_z \left| \frac{1}{N-1} \sum_{j=1}^{N-1} \in (Z - R_j^*) - G_0^*(z) \right| > d(\alpha, N-1).$$

This corresponds to a goodness-of-fit test for the modified radii $\{R_j^*\}$. One could have used any one of a number of other goodness-of-fit statistics, e.g., Cramer-von Mises. However, in this work, one uses consistently the Kolmogorov-Smirnov statistics.

The final detection problem to be considered here is a two-sample problem.

IX. SIGNAL DETECTION: A TWO-SAMPLE PROBLEM

$$\underline{PN_5}: \ L_1 = L_2 \in \Omega(K-S-E) \ \text{ vs } \ \underline{N + S_5}: \ L_1 \ne L_2$$

The data matrix here is

$$\tilde{Z} = \begin{pmatrix} X_{11} & \cdot \ \cdot \ \cdot & X_{1k} \\ \cdot & & \\ \cdot & & \\ \cdot & & \\ X_{m1} & \cdot \ \cdot \ \cdot & X_{mk} \\ & & \\ Y_{11} & \cdot \ \cdot \ \cdot & Y_{1k} \\ \cdot & & \\ \cdot & & \\ \cdot & & \\ Y_{m1} & \cdot \ \cdot \ \cdot & Y_{nk} \end{pmatrix}$$

where $N = m + n$.

The relevant statistics are

$$R_j^2 = \begin{cases} \sum_{s=1}^{k} X_{js}^2 & \text{for } 1 \le j \le m; \text{ and} \\ \sum_{s=1}^{k} Y_{js}^2 & \text{for } m+1 \le j \le N. \end{cases}$$

One now has

Decision Rule 9.1. Decide $N + S_5$ iff $\sum_1^m R(R_j) \geq a_1$ or $\leq a_2$.
[The table here is that for the Mann-Whitney-Wilcoxon Rank-Sum Statistic.]

Decision Rule 9.2. Decide $N + S_5$ iff

$$D(m, n) = \sup_z \left| \frac{1}{m} \sum_1^m \in (z - R_j) - \frac{1}{n} \sum_{m+1}^N \in (z - R_j) \right| > d'(\alpha, m, n)$$

[The table here is that of the two sample K-S Statistic.]

Again a large number of nonparametric (NP) statistics are available for this detection problem. The two chosen above are representative and have some optimal properties.

The preceding developments lead to some interesting observations and speculations.

X. CONCLUDING REMARKS AND OPEN PROBLEMS

(A) Parametric and NP Statistics

Signal detection problems related to SE time series utilize parametric as well as NP techniques. The sphericity property allows one to employ the F-statistic and other classical statistics; while the families of radial distributions are NP in character. Hence, one could have used a variety of goodness-of-fit and NP statistics. The relative utility of these other statistics is at this time an open question. [See (B) below.]

(B) FDR and Power

The procedures given in the text are reasonable and commonly used. However, very little precise information is available on FDR and power. A series of definitive studies, evaluating FDR for reasonable $(N + S)$-distributions, is needed. Some preliminary Monte Carlo results have been relatively costly and less than definitive. As of this writing, it is difficult to say which of the available procedures is better. In particular, one asks how do the procedures based on the Z's compare with those based on the Y's in Section 6.

(C) Polynomial Drift

Techniques were presented above for time series S-E about 0 and about non-zero points "a". Of some interest would be background noise which is SE about some polynomial curve. This would be related to polynomial regression with SE (rather than Gaussian) errors and should be also related to some recent robustness studies.

(D) Gaussian Markov Processes

It is developed that S-E time series can be considered "imbedded" in certain mixtures of some Gaussian Markov processes. It is an open problem as to how the very rich literature on Gaussian Markov processes can be brought to bear on signal detection problems.

REFERENCES

1. R. Ahmad, "Extension of the normal theory to spherical families," *Trab. de Estad.*, Vol. XXIII, pp. 51-60, 1972.

2. I.V. Basawa and B.L.S.P. Rao, "Statistical Inferences for Stochastic Processes," Academic Press, New York, 1980.

3. C.B. Bell, "Some basic theorems of distribution-free statistics," *Ann. Math. Statistics,* Vol. 55, pp. 150-156, 1964.

4. C.B. Bell, "Circularidad en Estadisticas," *Trabajos de Estadisticas,* Vol. XXVI, pp. 61-81, 1975.

5. C.B. Bell, "Statistical Inference for special families of Stochastic Processes," in Statistical Inference and Related Topics, ed. by Madan L. Puri, Academic Press, pp. 273-290, 1975.

6. C.B. Bell, Inference for goodness-of-fit problems with nuisance parameters, *J. Stat. Plan,* Inf. 9, pp. 273-284, 1984.

7. C.B. Bell, "Algunos Métodos Generales para la Construcción de Tests Paramétricos y No-parametricos," *Trab. Estad. Inv. Operativa,* Vol. 31, pp. 21-44, 1985.

8. C.B. Bell and K.A. Doksum, "Some new distribution-free statistics," *Ann. Math. Statistics,* Vol. 35, pp. 203-214, 1965.

9. C.B. Bell and Andrew L. Mason, "Nuisance Parameters, Goodness-of-fit tests and Kolmogorov-type statistics," *Proceedings of 1984 Conference on Goodness-of-fit,* Hungarian Academy of Sciences,1986.

10. C.B. Bell, F. Ramirez, and E. Smith, "Wiener-Levy Models, Spherical Exchangeable Time Series, and Simulataneous Inference in Growth Curve Analysis," Chapter 5 in Advanced Asymptotic Testing and Estimation: A Symposium in honor of Wassily Hoeffding, ed. by I.M. Chakravarti, Cambridge University Press, 1980.

11. C.B. Bell and P.K. Sen, Randomization procedures, in HandbookofStatistics, Vol. 4, ed. by P.R. Krishnaiah and P.K. Sen, North Holland Press, 1984.

12. C.B. Bell and P.J. Smith, "Some aspects of the concept of symmetry in Nonparametric Statistics," Symposium on SymmetricFunctionsinStatistics, honoring Paul Dwyer, ed. by Derrick S. Tracy, University of Windsor, pp. 143-181, 1972.

13. C.B. Bell, M. Woodroofe, and T.B. Avadhani, "Some nonparametric tests for stochastic processes," in Nonparametric Techniques in Statistical Inference, ed. by Madan L. Puri, Cambridge University Press, pp. 215-258, 1970.

14. Y.J. Choi, "Kolmogorov-Smirov Test with Nuisance Parameters in the Uniform Case," Master of Science thesis, University of Washington, 1980.

15. J. Durbin, "Some methods of constructing exact tests," *Biometrika,* Vol. 48, pp. 41-55, 1961. Correction 1966, *Biometrika,* Vol. 53, p. 629.

16. C.R. Rao, S.K. Mitra, and A. Matthai (eds), Formulae and Tables for Statistical Work, Statistical Publishing Society, Calcutta, India, Section 10, Nonparametric Tests, Table 10.1.

17. B. Efron and R.A. Olshen, "How broad is the class of normal scale mixtures?" *Ann. Statist.,* Vol. 6, pp. 1159-1164, 1978.

18. D. Kelker, "Distribution theory of spherical distributions and a location-scale parameter generalization," *Sankhya, Ser. A.,* Vol. 32, pp. 419-430, 1970.

19. D. Kelker, "Infinite divisibility and variance mixtures of the normal distribution," *Ann. Math. Statistic,* Vol. 42, pp. 824-827, 1971.

20. A.N. Kolmogorov, "Sulla determinazione empirica di una legge di distribuzione," *G. Ist. Ital. Attuari,* Vol. 4, p. 83, 1933.

21. M.L. King, "Robust test for spherical symmetry and their application to least squares regression," *Ann. of Statistics,* Vol. 8, pp. 1265-1272, 1980.

22. H.W. Lilliefors, "On the Kolmogorov-Smirnov test for normality with mean and variance unknown," *JASA,* Vol. 62, pp. 399-402, 1967.

23. H.W. Lilliefors, "On the Kolmogorov-Smirnov test for the exponential distribution with mean unknown," *JASA,* Vol. 64, pp. 387-389, 1969.

24. R.D. Lord, "The use of Hankel transformation in statistics, I. General Theory and examples," *Biometrika,* Vol. 41, pp. 44-55, 1954.

25. A.L. Mason, "Kolmogorov-Smirnov Type Statistics for Goodness-of-fit Problems with Nuisance Parameters," *M.S. Project,* San Diego State University, 1984.

26. A.L. Mason and C.B. Bell, "New Lilliefors and Srinivasan Tables with Applications," *Commun. Statist.-Simula.,* Vol. 15, No. 2, pp. 451-477, 1986.

27. E.J. Schoenberg, "Metric spaces and completely monotone functions," *Ann. Math.,* Vol. 39, pp. 411-841, 1938.

28. R. Srinivasan, "An approach to testing the goodness-of-fit of incompletely specified distributions," *Biometrika,* Vol. 57(3), pp. 605-611, 1970.

29. P.J. Smith, Structure of Nonparametric Tests of some Multivariate Hypothesis, Ph.D. Thesis, Case Western University, 1969.

CONTRIBUTIONS TO NON-GAUSSIAN SIGNAL PROCESSING

Roger F. Dwyer
Naval Underwater Systems Center
New London, Connecticut 06320

ABSTRACT

Data from the 1980 and 1982 Arctic experiments and from in air helicopter-radiated noise measurements are presented and discussed as part of the physical evidence for the existence of non-Gaussian signals and noises. After the data are presented, theoretical considerations are given for frequency domain kurtosis estimation. Statistical models are introduced and discussed which support and correspond to the measurements made on the real data in the frequency domain. It was discovered from frequency domain measurements that at times Arctic under-ice noise consists of narrowband non-Gaussian interference components. These components can severely degrade the performance of sonar systems. Therefore, a method was developed that effectively removes narrowband interference by implementing a non-linearity in the frequency domain.

I. INTRODUCTION

From the likelihood ratio (LR) criterion, detection performance depends on a ratio of the probability distribution of signal and noise to the probability distribution of noise only. When the probability distributions are not Gaussian, signal detection and estimation performance can usually be improved based on the LR procedure over systems which implicitly or explicitly utilize the Gaussian assumption. The first definitive work on non-Gaussian noise problems can probably be attributed to Capon [1]. Kurz [2] introduced non-parametric methods based on quantiles for detection and estimation problems. In 1976 Dwyer [3] introduced non-Gaussian sequential detection methods based on quantiles for application to sonar problems. This work on partitioning also included data with Markov dependency [3]. Subsequently, these results were generalized to multidimensional Markov processes and applied to array signal detection problems [4].

Other significant contributions to the field of non-parametric signal detection were made by Thomas and his students [5,6]. In addition, robust methods based on Huber's [7] work were contributed by Martin and Schwartz [8]. In these studies the physical existence and occurrence of non-Gaussian noise were not considered. However, Middleton discused in a general setting the physical significance of non-Gaussian noise and he subsequently developed statistical physical models of non-Gaussian noise environments [9]. Although the physical existence of non-Gaussian noise and interference now are well documented in electromagnetic communication systems, supporting evidence was lacking in the underwater acoustic environment. Therefore, the major goals of our part of the non-Gaussian signal processing program were the documentation of non-Gaussian signals and noises related to sonar systems and the measurement of performance improvements based on the theoretical concepts already developed. However, real underwater acoustic measurements revealed unexpected results which elicited new theoretical concepts that became an important contribution to the program.

Two data sets were analyzed. Data collected during the 1980 and 1982 [10] Arctic experiments were analyzed for non-Gaussian noise by using time domain and frequency domain techniques [11]. Based on the frequency domain analyses, it was revealed that narrowband interference existed in the Arctic data. The mechanisms for the narrowband interference are believed to be due to frictional interactions and colliding ice floes. It was obvious that these interferences would seriously degrade the performance of sonar systems. Therefore, an investigation was begun to develop methods to reduce the effect of ice induced narrowband interference. The other data set was from helicopter-radiated noise. It was discovered that impulsive noise, due to blade vortex interaction, was measurable in the time and frequency domains. A frequency domain kurtosis (FDK) estimate [11] showed significantly high FDK levels in the frequency domain corresponding in many cases to the discrete radiated frequencies. In the analysis of these real data sets the mean, variance, skew, and kurtosis were estimated for the real and imaginary parts of each frequency component. Based on the measurements, the kurtosis estimate appeared to be more significant and a theoretical analysis was, therefore, undertaken. These results will be discussed later in this paper.

II. ARCTIC UNDER-ICE AMBIENT NOISE AND PROCESSING

As stated above, the data analyses are composed of time and frequency domain statistical measurements. The time domain data were filtered, sampled, and grouped into records of 1024 samples each. The mean, variance, skew, and kurtosis were then estimated for each record. Over time intervals consisting of hundreds of records, the cumulative distribution function (CDF) was estimated and was shown, for the most part, to be non-Gaussian, but with nonstationary behavior over successive intervals. These results were consistent with the results of Milne and Ganton who made measurements in the Canadian Arctic Archipelago [12].

The time domain data were then transformed into the frequency domain via a 1024 point fast Fourier transform (FFT). Frequency domain statistics were then compiled for each frequency cell for both the real and imaginary parts. It was discovered that the frequency domain contained strong non-Gaussian noise. This significant new discovery has important implications for sonar performance improvements. Figure 1 gives an example of the kurtosis spectrum and compares it with the classical power spectrum. The top curve in the figure represents the power spectrum. On the left, is an example of a 60-Hz tonal due to an electrical ground loop. The corresponding FDK estimate indicated in the bottom curve by a dip which has a value of 1.8. Other dips in the figure are due to harmonics of the 60-Hz ground loop. This result was predicted in a theoretical analysis which will be summarized later. As is clear from the bottom curve, many frequencies deviate significantly from a Gaussian process based on the FDK estimate.

Recall from the likelihood ratio formulation that an optimum detection statistic utilizes knowledge of the signal and noise and noise only probability distributions. Therefore, in non-Gaussian noise environments sonar performance may be improved by taking into account the underlying environmental noise probability distribution. One goal of our research was to document the potential for performance improvements using the Arctic under-ice data. Using an adaptive partitioning method [3], which approximates the optimum nonlinearity, in the frequency domain, it was shown that for the under-ice environmental noise, detection threshold reduction up to 12 dB and noise variance reductions of 5 dB could be achieved.

The non-Gaussian noise in the frequency domain is believed to be due to rubbing ice floes. This type of interference can also significantly degrade the performance of systems which estimate correlation functions to obtain bearing and range information. It was found that the FDK estimate could distinguish between the desired signal and the unwanted ice sounds. A method was developed for removing the ice induced narrowband frequency components. It consisted of passing the frequency domain ice noise data through a nonlinearity, and then transforming back into the time domain. Examples showed that narrowband interference could be removed efficiently from the information containing signal and, therefore, range information could be improved.

III. HELICOPTER-RADIATED NOISE

Another important data set collected for the verification of non-Gaussian processes was helicopter radiated noise. In air measurements were made by Dwyer of a UH-1 helicopter flying at an altitude of approximately 600m (2,000 ft.). Essentially, the same statistical measurements were made for the in air helicopter-radiated noise as were for the under-ice ambient noise.

Time domain data show that radiated noise of a UH-1 helicopter consists of measurable impulsive noise due to blade vortex interactions. Figure 2 compares the time domain cumulative distribution function of the in air radiated noise measurements with Gaussian noise. The in air data clearly show that the CDF is non-Gaussian in both the high probability and low probability regions. Whereas, the in air data with additive Gaussian noise only appears to be non-Gaussian in the tails of the distribution. This result would be expected in noisy environments. But the in air data clearly show the impulses from the blade vortex interaction and their periodicity can be measured. For the in air data, autocorrelation function estimates show autocorrelation values around .8 which occur periodically with nearly the same value with increasing time delay. The reciprocal of the time interval between impulses also correspond to the fundamental frequency of the radiated noise.

Therefore, as expected, the frequency domain results showed that the in air radiated noise from a UH-1 helicopter is composed of frequency harmonics from the main rotor and tail rotor measurable out to about 400 Hz. The kurtosis spectrum reveals that many frequency components are associated with high kurtosis levels. However, some high kurtosis levels are not associated with discrete power spectrum components. The reason for this is that the kurtosis estimate is sensitive to the probability distribution of the radiated noise, whereas, the power spectrum is sensitive to its energy content.

The conclusions reached by the statistical analyses of the in air helicopter-radiated noise measurements are as follows:

1. Helicopter (UH-1) radiated noise can be detected in the frequency domain by its narrowband components up to 400 Hz.
2. High-order spectral moments (kurtosis spectrum) indicate unsteady radiated frequency components.
3. Broadband radiated noise levels are measurable in the time and frequency domains.
4. The estimated cumulative distribution function in the time domain exhibit non-Gaussian characteristics.

IV. THEORETICAL ASPECTS OF FREQUENCY DOMAIN KURTOSIS ESTIMATION

In the frequency domain analysis of the real data discussed above, the mean, variance, skew, and kurtosis were estimated for the real and imaginary parts of each frequency component. The kurtosis estimate appeared to be more significant and a theoretical analysis was, therefore, undertaken. From the theoretical work, frequency domain kurtosis (FDK) estimation can be considered as a generalization of power spectrum density (PSD) analysis. This is true because the PSD is essentially a sum of the estimates of the second order moments for both the real and imaginary parts for each frequency component in the frequency domain. If the frequency domain signals are not Gaussian distributed, then higher order moments of the complex frequency components would contain additional information that may be utilized in signal processing. The objective of the theoretical analysis was to compare the PSD technique for signal processing with the new method which estimates frequency domain kurtosis for the real and imaginary parts of the complex frequency components.

In order to better understand the detection of randomly occurring signals, sinusoidal and narrowband Gaussian signals were considered, which when modeled to represent a fading or multipath environment, were detected as non-Gaussian in terms of a frequency domain kurtosis estimate. Several fading and multipath propagation probability density distributions of practical interest were considered, including Rayleigh and log-normal. The model was also generalized to handle transient and frequency modulated signals by taking into account the probability of the signal being in a specific frequency range over the total data interval. It was shown that this model produces kurtosis values consistent with the real data discussed previously. An important theoretical result was obtained by comparing the ability of the power spectral density estimate and the frequency domain kurtosis estimate to detect randomly occurring signals, generated from the model, using the deflection criterion. It was shown, for the cases considered, that the frequency domain kurtosis estimate for a limited range of conditions had a processing advantage.

The measure for comparison was the deflection criterion. According to Lawson and Uhlenbeck, a detectability criterion can be defined in terms of a function fo deflection. The detectability criterion is defined as a ratio of the difference between the average value of a function when signal and noise are present and the averaged value of the function when noise only is present, to the standard deviation of the function under noise only conditions. The function is either the FDK or the PSD. The ratio of the two criteria for the FDK and PSD defines a relative efficiency for the two methods. Figure 3 shows the relative efficiency vs. the probability of occurrence of the signal which is parameterized by L. The parameter A_1 represents a signal-to-noise ratio measurement. The figure shows that for small values of the parameter L, the relative efficiency will be greater than 1 and therefore indicates that the FDK is a better detection statistic than the PSD. But, even in those cases where the PSD is a better detection statistic, the FDK can still be useful. For example, the physical evidence of the real data suggests the existence of measurable random occurring signals and, therefore, the FDK estimate may provide additional information for detecting and classifying those signals.

In addition to the deflection criterion, the probability of detecting a non-Gaussian signal for a fixed false alarm rate was analytically derived for both the FDK and PSD estimates. This derivation represented a new analytical approach for comparing complicated statistics. The method was based on Cramer's convergence proof that higher-order statistics, which includes kurtosis and second-order estimates, converged to a Gaussian process. Therefore, only the mean and variance were needed in the probability of detection comparison. Figure 4 shows a representative case of the performance comparison between the FDK and PSD estimates. The figure compares the asymptotic probability of detection (APD) vs. the probability that the signal occurs in the interval. The solid curve represents the FDK and the broken curve the PSD estimate. The results are shown as a function of signal-to-noise ratio (SNR). For a fixed SNR the asymptotic probability of detection for the FDK will increase as L increases up to a point, then decrease. The reason for this is that the FDK is sensitive to non-Gaussian signals and for $L = 1$ the signal is Gaussian so detection is impossible. The figure shows that for small values of L the FDK can have a higher asymptotic probability of detection compared with the PSD. Therefore, these results could be important in applications where it is desired to detect a non-Gaussian signal. The above results have also been extended to cases where a noise only sample is available. In this case, the modified FDK has a higher asymptotic probability of detection level compared with the PSD under all conditions considered.

V. IDENTIFICATION AND REMOVAL OF NARROWBAND INTERFERENCE

A method to remove narrowband interference in the frequency domain was developed. It consists of first transforming the data into the frequency domain by an FFT, passing the transformed data through an ideal nonlinearity, and then transforming the data back into the time domain by an IFFT. The essential mathematical details of the method are given here and error criteria are defined which measures the effectiveness of the technique.

As pointed out above, Arctic under-ice noise is at times composed of narrowband components [13]. The narrowband noise is primarily due to rubbing ice flows but possibly acoustic dispersion contributes to this phenomenon. This type of interference can significantly degrade the performance of systems which estimate autocorrelation functions to obtain bearing and range information. A typical sample of Arctic under-ice narrowband interference is shown in Figure 5. The data below 2 kHz is broadband noise. But a narrowband component is present in the figure above 2 kHz which lasts for 12.8 seconds.

Many segments of Arctic under-ice data contained these highly dynamic narrowband components as shown in Figure 5. The statistical behavior of the dynamic narrowband frequency components were measured by first transforming the data into the frequency domain using a fast Fourier transform (FFT). Then, as discussed previously, kurtosis was estimated for each real and imaginary part of each frequency component over the band for a group of consecutive FFT segments. Thus, the FDK estimates the distribution over a time interval consisting of many FFT segments for each real and imaginary frequency component. Many of the Arctic data segments showed non-Gaussian components due to the highly dynamic nature of the narrowband ice components. Therefore, the FDK is a method whereby a stable signal can be distinguished from the unwanted ice sounds.

Once the narrowband interference is identified it can be removed by passing the data of that frequency component through a nonlinearity [13]. The output data with the interfering component removed may then be transformed back into the time domain for further processing. For example, the autocorrelation function of the desired signal may be estimated free of the interfering narrowband noise. Or, for two channels, the cross-correlation function may be estimated after both channels are processed through the nonlinearity in the frequency domain. The smoothed coherence transform (SCOT) introduced by Carter, Nuttall, and Cable [14], is a technique which improves time delay estimation between broadband correlated signals in the presence of strong narrowband interference. The SCOT utilizes a frequency domain whitening process of the cross-spectrum. The SCOT processed data is then transformed into the correlation domain so that time delay can be estimated. Hassab and Baucher [15] have utilized other window functions to improve SCOT performance for signal and noise with smooth spectra.

In contrast to the SCOT and its generalizations, our method is applied to one channel, or in multichannel cases, to each channel separately. Also, the narrowband interference is removed by passing the frequency domain data through a nonlinearity in contrast to whitening. Therefore, only those frequency components that are deemed interference by the FDK estimate are removed. In addition, the optimum nonlinearity can be derived from a likelihood ratio formulation under the assumption of independent observations [13]. The resultant data are in the time domain with the interference removed. The removed data are also available as a function of time. This allows performance criteria to be defined in the time domain.

Time domain techniques may also be employed. For example, data adaptive signal estimation by singular decomposition has been proposed and evaluated in reference 16, which could be used to estimate and remove the interfering components.

Next, an ideal nonlinearity (INL) will be utilized so that the essential mathematical features of the method can be discussed conveniently. However, the INL is similar to the nonlinearity discussed in reference 13.

It will also be clear from the following development that the method can also be applied to the spatial domain as well as the frequency domain. But, the discussion will only be concerned with the frequency domain.

THE IDEAL NONLINEARITY

Let $x(i)$, $i = 0, 1, 2, \cdots, N-1$, represent the real discrete data. The discrete Fourier transform (DFT) of $x(i)$ is

$$X(k) = \sqrt{1/N} \sum_{i=0}^{N-1} w(i)\, x(i)\, exp(-j2\pi ki/N) \quad ,$$

where $j = \sqrt{-1}$, and $k = 0, 1, 2, \cdots, N-1$.

For simplicity, the window weights are set equal to one, i.e., $w(i) = 1$ for all i.

It can be shown that the components are related by the relationship, $X(k) = X^*(N-k)$, for $k = 0, 1, 2, \cdots, N$, with $X(0) = X(N)$, since $X(0)$ and $X(N)$ are real. The asterisk represents complex conjugate.

If the input data are an additive mixture of signal, noise, and interference of the form,

$$x(i) = s(i) + n(i) + I(i) \, ,$$

then the components in the frequency domain are,

$$X(k) = S(k) + N(k) + I(k) \, .$$

The signal, $s(i)$, which is the information bearing component of the received data, is corrupted by noise, $n(i)$, and interference $I(i)$.

If the interference is narrowband and within the bandwidth of the signal it will generally degrade the autocorrelation estimate of the signal. The approach that will be considered to rectify this problem is to remove the interfering components from the signal by using an INL.

The INL is defined for complex values, as

$$X(k) = \mathbb{0}, \text{ if } k = kg + md$$

$$X(k) = \mathbb{0}, \text{ if } k = N - (kg + md)$$

$$X(k) = X(k), \text{ otherwise },$$

where $m = 0, 1, 2, \cdots, I-1$, d = integer constant, and $0 \leq kg + md \leq N/2$, and $N/2 \leq N - (kg + md) \leq N$. Here, the symbol $\mathbb{0}$, means that both real and imaginary parts are set to zero.

The interfering frequencies start at kg and extend to $kg + (I-1)d$. In order to include the interference that may be periodic in the frequency domain the parameter, d, will not equal one, i.e., $d \neq 1$.

For example 60 Hz interference and its harmonically related frequencies are sometimes present in systems. In this case d may represent the number of frequency bins between components. Also, the periodic frequency domain components of helicopter-radiated noise may be removed with this method.

Once the interfering components have been identified and removed by the INL, the output is given by

$$Y(k) = X(k) - \sum_{m=0}^{I-1} X(k)\,\{\delta[k - (kg + md)] + \delta[k - N + (kg + md)]\}$$

where the Kronecker delta function $\delta(k-p)$ is equal to one when $k = p$ and is equal to zero otherwise.

The inverse DFT of $Y(k)$ is

$$y(i) = s(i) + n(i) + I(i) - I_R(i),$$

where

$$I_R(i) = \sqrt{1/N} \sum_{m=0}^{I-1} \{X(kg + md)\, exp\,[j2\pi i\,(kg + md)/N] + X^*(kg + md)\, exp\,[-j2\pi i\,(kg + md)/N]\} .$$

The function $I_R(i)$ can be interpreted as an inverse Fourier transform over the frequencies, kg through $kg + (I-1)d$, which are removed from the input $x(i)$.

Therefore, all the frequencies contained in $I_R(i)$ will be removed from $x(i)$. If the signal is separated, in frequency space, from $I_R(i)$, then only the interference and those frequencies of the noise contained in $I_R(i)$ will be removed. However, if part of the signal is contained in the interfering frequency space it will also be removed. This represents a disadvantage of the method. In some applications the partial loss of signal may be tolerated if the overall performance is improved.

The effectiveness of the INL may be measured from the error equation

$$\text{Signal Error} = \overline{s(i)I_R(i)}/\overline{s(i)^2} ,$$

where $s(i)$, $n(i)$, and $I(i)$ are assumed mutually independent and zero mean processes.

In addition, an error equation can also be defined for the interference as follows:

$$\text{Interference Error} = 1 - \overline{I(i)\, I_R(i)}/\overline{I(i)^2}$$

In applications, the objective would be to minimize both errors and these would define performance criteria for judging the effectiveness of the method.

EXAMPLES

1. Consider a case where the data are two pure sinusoids at two different frequencies. In addition, for the FFT results, assume the frequencies are centered in the frequency bins. One sinusoid represents the signal and the other interference. Therefore, the error will be zero. Figure 6 shows the sum of the two sinusoids in the top graph. The interference is obscuring the signal in the time domain. The bottom graph shows the signal after the interference was removed by the method of using a nonlinearity in the frequency domain.

2. The last example concerns measuring the autocorrelation function of a broadband Gaussian signal. However, a strong additive sinusoid is present in the data. Figure 7 shows the time history of the additive broadband signal and interference in the top graph. The bottom graph represents the time history after the sinusoid has been removed. Since the signal was not completely disjoint from the interference, in frequency space, part of the signal corresponding to the interfering frequency was also removed. Therefore, the signal error was not zero but, nevertheless, small. The benefit from the method is obvious from the figure. This result is probably appreciated more by observing the difference in the autocorrelation function estimate. Figure 8 represents the autocorrelation function estimate of signal and interference in the top graph. The interference has completely dominated the estimate. The bottom graph shows the same estimate after the interference has been removed.

VI. SUMMARY

During the three years of the non-Gaussian signal processing program, the author's efforts were initially devoted to obtaining physical evidence for the existence of non-Gaussian signals and noises. As a consequence of these successful efforts, underwater acoustic measurements revealed unexpected results which elicited new theoretical concepts and methods to extract useful information and remove narrowband interference.

The data analyzed consisted of the 1980 and 1982 Arctic under-ice ambient recordings and in air helicopter-radiated noise measurements made by the author. It was discovered that the under-ice data contained narrowband non-Gaussian noise measurable in the frequency domain and impulsive noise measurable in the time domain. The latter data were consistent with the results of Milne and Ganton who made measurements in the Canadian Arctic Archipelago. However, the frequency domain narrowband results were new. It is believed that the mechanism for the narrowband components are due to frictional interactions of colliding ice flows. Since it was obvious that this kind of interference would seriously degrade the performance of sonar systems, a statistical model was developed to predict the measured results. In addition, a method was developed to remove these components in the frequency domain and thereby improve sonar system performance in these environments.

The in air helicopter-radiated noise measurements also consisted of time domain impulsive noise from blade vortex interactions and non-Gaussian noise in the frequency domain mainly due to frequency modulation produced by doppler effects. The impulsive noise was clearly periodic and autocorrelation function estimates revealed autocorrelation levels near .8 which occurred periodically with nearly the same level with increasing time delay.

For the frequency domain analysis, a new measurement was introduced. It consisted of estimating kurtosis at the output of a discrete Fourier transform for each real and imaginary part of each frequency component. This estimate revealed non-Gaussian interference in the frequency domain. From theoretical considerations frequency domain kurtosis (FDK) estimation was shown to be a generalization of power spectrum density (PSD) estimation in that the FDK is sensitive to the probability distribution of the data, whereas, the PSD is sensitive to its energy. Therefore, the FDK can contribute additional information that may be useful in signal processing applications.

Since narrowband interference can degrade systems which estimate correlation functions to obtain range and bearing information, a method was developed to remove it. It consisted of passing the transformed data through a nonlinearity and transforming the resultant data back into the time domain. Examples showed that this method was very effective in removing narrowband interference.

REFERENCES

1. J. Capon, "On the Asymptotic Efficiency of Locally Optimum Detectors," *IRE Trans. on Inf. Theory,* Vol. IT-17, No. 1, pp. 67-71, (1961).

2. Y. Ching and L. Kurz, "Nonparametric Detectors Based on M-Interval Partitioning," *IEEE Trans. on Inf. Theory,* Vol. IT-18, pp. 251-257, (1972).

3. R. Dwyer, "Sequential Partition Detectors with Applications," Ph.D. Dissertation, Polytechnic Institute of New York, 1976.

4. R. Dwyer, "Theory of Partition Detectors for Data Represented by Markov Chains," *NUSC Technical Memorandum,* TM 811031, 2 April 1981.

5. J. Carlyle and J. Thomas, "On Nonparametric Signal Detectors," *IEEE Trans. on Inform. Theory,* Vol. IT-10, pp. 146-152, April 1964.

6. J. Miller and J. Thomas, "Detectors for Discrete-Time Signals in Non-Gaussian Noise," *IEEE Trans. on Inform. Theory,* Vol. IT-18, pp. 241-250, 1972.

7. P. Huber, "Robust Estimation of a Location Parameter," *Ann. Math. Statist.,* Vol. 35, pp. 73-101, 1964.

8. R. Martin and S. Schwartz, "Robust Detection of a Known Signal in Nearly Gaussian Noise," *IEEE Trans. on Inform. Theory,* Vol. IT-17(1), pp. 50-56, 1971.

9. A. Spaulding and D. Middleton, "Optimum Reception in an Impulsive Interference Environment - Part I: Coherent Detection," *IEEE Trans. on Comm.,* Vol. COM-25, No. 7, pp. 910-923, 1977.

10. F.R. DiNapoli, et.al., "Tristen/FRAM II Cruise Report East Arctic, April 1980," *NUSC,* TD 6457, 13 April 1981.

11. R.F. Dwyer, "FRAM II Single Channel Ambient Noise Statistics," *NUSC,* TD 6583, 25 November 1981.

12. A.R. Milne and J.H. Ganton, "Ambient Noise Under Arctic Sea Ice," *Journ. Acoust. Soc. Am.,* Vol. 36, pp. 855-863, 1964.

13. R. Dwyer, "A Technique for Improving Detection and Estimation of Signals Contaminated by Under Ice Noise," *J. Acoust. So. Am.,* Vol. 74(1), July 1983.

14. G.C. Carter, A.H. Nuttall, and P.G. Cable, "The Smoothed Coherent Transform," *Proceedings of the IEEE (Lett),* Vol. 61, pp. 1497-1498, October 1973.

15. J. Hassab and R. Boucher, "Performance of the Generalized Cross Correlator in the Presence of a Strong Spectral Peak in the Signal," *IEEE Trans. on ASSP,* Vol. ASSP-29, No. 3, June 1981.

16. D. Tufts, R. Kumaresan, and I. Kirsteins, "Data Adaptive Signal Estimation by Singular Value Decomposition of a Data Matrix," *Proceedings of the IEEE,* Vol. 70, No. 6, June 1982.

LIST OF PUBLICATIONS

1. R. Dwyer, "Acoustic Problems," *Papers presented at the SRO II Workshop held at NUSC,* p. 20 in NUSC TD 6591, 7-8 April 1981.

2. R. Dwyer, "FRAM II Single Channel Ambient Noise Statistics," presented at the *101st Meeting of the Acoustical Society of America,* Ottawa, Canada, 18-22 May 1981. Published in *NUSC Tech. Doc. 6583,* 25 November 1981.

3. R. Dwyer, "Arctic Ambient Noise Statistical Measurement Results and Their Implications to Sonar Performance Improvements," in *Proceedings of the Conference on Undersea Ambient Noise,* SACLANT ASW Research Centre, La Spezia, 11-14 May 1982.

4. R. Dwyer, "A Statistical Frequency Domain Signal Processing Method," in *Proceedings of the 16th Annual Conference on Information Sciences and Systems,* held at

Princeton University, 17-19 March 1982. NUSC reprint Rep. 6687, 22 March 1982.

5. R. Dwyer, "Detection of Non-Gaussian Signals by Frequency Domain Kurtosis Estimation," *IEEE IC-ASSP-83,* pp. 607-610, Conference Record.

6. R. Dwyer, "An Adaptive Method to Improve Signal Detectability in Non-Gaussian Noise Environments," *Third Yale Workshop on Applications of Adaptive System Theory,* Yale University, Vol. 3, pp. 68-73, 1983.

7. R. Dwyer, "A Technique for Improving Detection and Estimation of Signals Contaminated by Under Ice Noise," *Journ. Acoust. Soc. Am.,* Vol. 74, No. 1, pp. 124-130, 1983.

8. R. Dwyer, "Use of the Kurtosis Statistic in the Frequency Domain as an Aid in Detecting Random Signals," *IEEE Journal of Oceanic Engineering,* Vol. OE-9, No. 2, pp. 85-92, April 1984.

9. R. Dwyer, "Essential Limitations to Signal Detection and Estimation: An Application to the Arctic Under Ice Environmental Noise Problem," *IEEE Proceedings (Lett),* Vol. 72, No. 11, November 1984.

10. R. Dwyer, "Detection of Randomly Occurring Signals Using Spectra and Frequency Domain Kurtosis Estimates," *Naval Underwater Systems Center (NUSC) Technical Memorandum,* TM 841057, 23 March 1984.

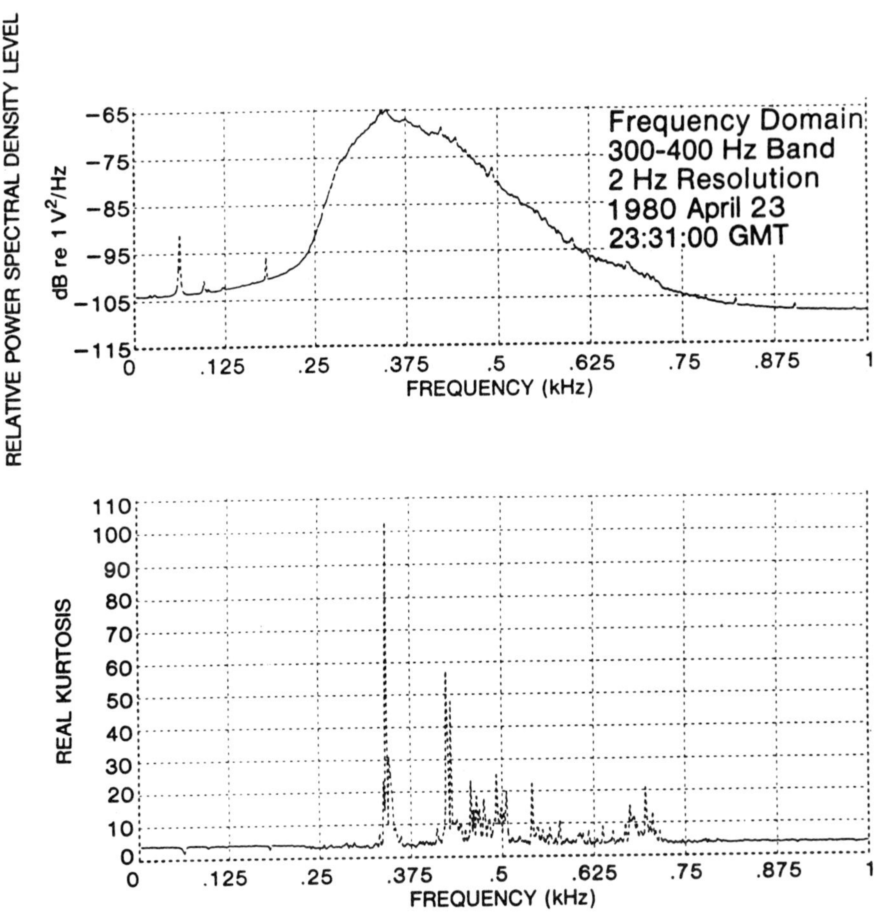

Figure 1 Power Spectrum and Frequency Domain Kurtosis Estimates vs Frequency.

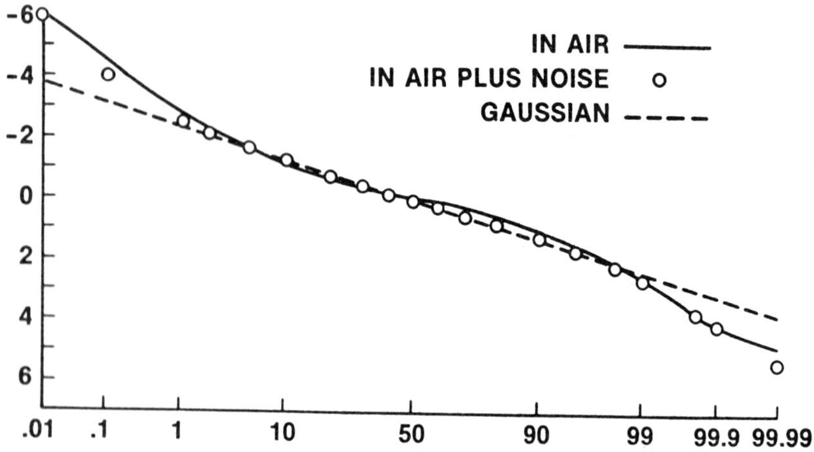

Figure 2 Cumulative Probability Distribution of Helicopter-Radiated Noise Measured in Air and with Additive Noise.

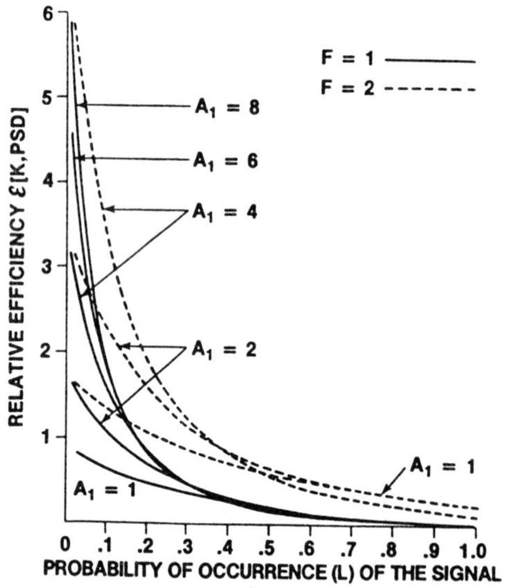

Figure 3 Relative Efficiency of FDK Relative to PSD vs Probability of Occurrence (L) of the Signal.

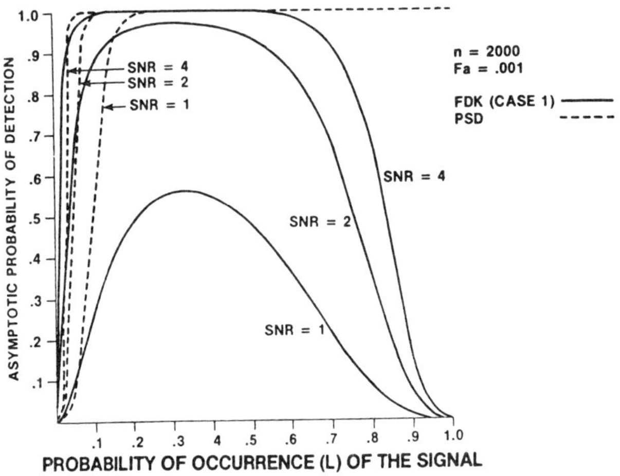

Figure 4 Asymptotic Probability of Detection vs. Probability of Occurrence (L) of the Signal.

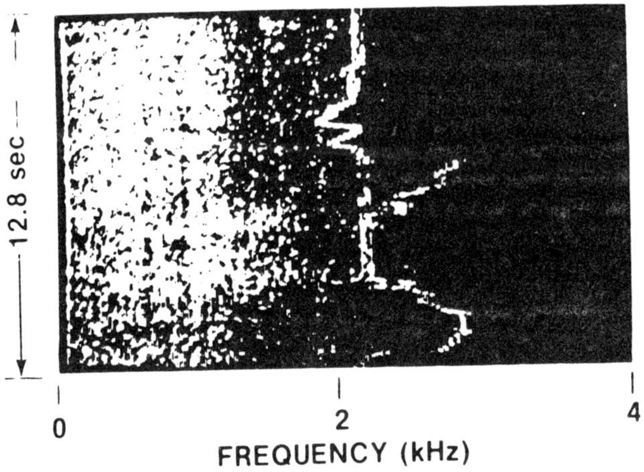

Figure 5 Frequency Domain Arctic Narrowband Interference.

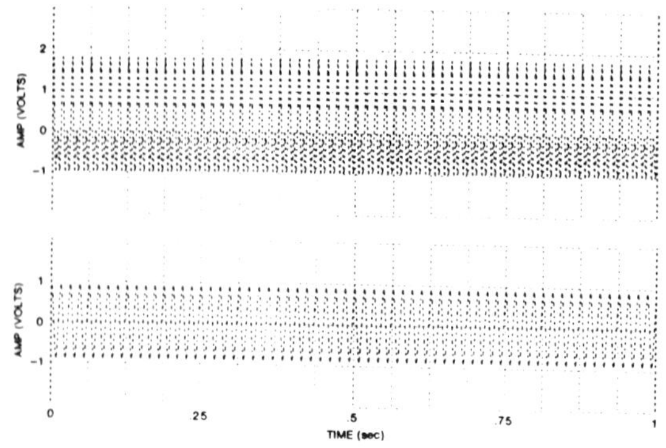

Figure 6 Sinusoidal Interference: Top, Interference and Signal; Bottom, Signal After Interference was Removed by a Nonlinearity.

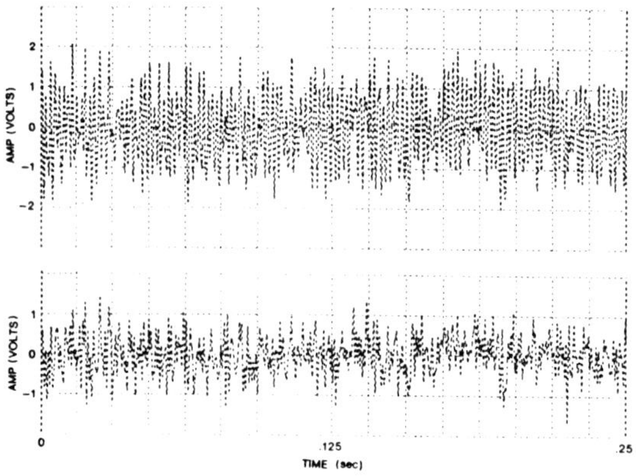

Figure 7 Broadband Signal with Sinusoidal Interference: Top, Time History of Signal and Noise; Bottom, Broadband Signal After Sinusoidal has been Removed by a Nonlinearity.

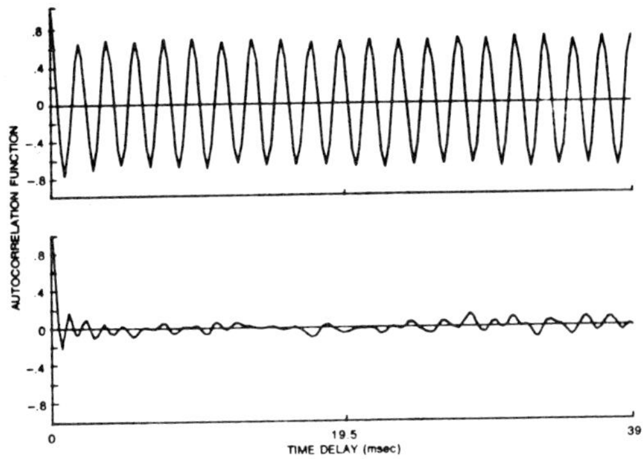

Figure 8 Autocorrelation Function: Top, Boradband Signal and Sinusoid; Bottom, Autocorrelation Function of Broadband Signal After Sinusoid has been Removed by a Nonlinearity.

DETECTION OF SIGNALS IN THE PRESENCE OF STRONG, SIGNAL-LIKE INTERFERENCE AND IMPULSE NOISE

D.W. Tufts, I.P. Kirsteins*, P.F. Swaszek,
A.J. Efron, and C.D. Melissinos
Department of Electrical Engineering
University of Rhode Island
Kingston, Rhode Island 02881

I. INTRODUCTION

We assume that the noise which interferes with signal detection can be considered to be a mixture of non-stationary, high-amplitude, non-Gaussian components plus a low amplitude Gaussian stationary component. Such a model appears to be widely applicable. The methodology that we propose for signal detection is to identify, categorize, model, and remove the non-Gaussian components in a piece-wise fashion based on their ease of separability from the background Gaussian noise and weak signals. This approach to modeling and processing complicated and non-stationary data is similar to that of experimental physicists going back at least to the time of Newton and perhaps most clearly articulated by Eugene Wigner in his Nobel Prize lecture [1]. Liu and Nolte [2] and Claus, Kadota, and Romain [3] have shown that when the noise is Gaussian and consists of a sum of a strong highly coherent component and a weak component of independent noise samples, then estimation and subtraction of the coherent noise component is nearly optimal. The application of special data smoothers and data cleaners by Martin and Thomson [4] for obtaining robust spectral estimates when the data is contaminated by outliers has provided another motivation for our use of adaptive differential quantization for robust detection.

Our proposed technique of iterative processing converts the problem of dealing with complicated non-stationary and non-Gaussian multivariate data to a sequence of simpler modeling problems. We have applied this methodology to processing a set of single channel digitized Arctic undersea acoustic data. We [6], Veitch and Wilk [5], and Dwyer [25] have studied this data set and hypothesize that it can be modeled as a mixture of three components as follows:

(1) weak stationary Gaussian background

(2) a number of strong non-stationary sinusoidal-like components which can occur randomly throughout the data

(3) sporadic high intensity impulsive bursts.

II. ESTIMATION OF PARAMETERS OF THE INTERFERENCE COMPONENTS AND REMOVAL OF THE INTERFERENCE

If a burst of interference can be modeled as an exponential signal, that is a combination of sinusoids or exponentially damped sinusoids, then we apply our previously developed principal component, low-rank estimation techniques [8-13]. The low-rank

* Now at the Naval Underwater Systems Center, New London, Connecticut.

method for improvement of SNR [8,6,21] and for detection [13] can be used even if the interference is not exponential. This implies that if we consider weak signal detectability in the presence of strong transients or tonals, we can estimate and remove the transient while leaving the weak signal essentially intact [13]. This method is not a magical panacea. If there is a significant portion of the signal in the interference subspace, then much of the signal will be removed also. We will return to the removal of coherent interference in Section 4. In this section and the next we concentrate on the removal of impulse noise.

First, we consider the robust removal of the impulses using the model

$$y_n = x_n + v_n \tag{1}$$

where y_n is the observed data, x_n is the smooth component of the data, consisting of signal, coherent interference, and background noise, and v_n are impulses or outliers. The rate of change of the smooth process x_n is defined as

$$r_n = x_n - x_{n-1} \tag{2}$$

The rate of change r_n can be viewed as an crude estimate of an innovations process. The adaptive differential quantization algorithm is given below:

First, robust estimates for the standard deviations of the smooth process x_n and the smooth-process rate of change r_n are obtained using median type estimators based on N previous samples as follows:

$$Dev[r_n] \approx \tilde{S}_n = \text{median}\{\,|\Delta_{n-N}|,\ |\Delta_{n-N+1}|,\ \cdots,\ |\Delta_{n-1}|\,\} \tag{3}$$

$$Dev[x_n] \approx \tilde{V}_n = \text{median}\{\,|y_{n-N}|,\ |y_{n-N+1}|,\ \cdots,\ |y_{n-1}|\,\} \tag{4}$$

where

$$\Delta_n = y_n - y_{n-1} \tag{5}$$

and we assume that x_n, y_n, and Δ_n have zero mean values.

The adaptive quantizer estimates an interval within which the next sample of the smooth component, x_n, of the data element y_n is very likely to lie if no impulse occurs with that sample. This interval is centered on an estimate of the conditional mean of x_n, and the width of the interval is based on an estimate of the conditional standard deviation of x_n. The estimates must be robust, because of the presence of the impulses.

These confidence intervals for the observed data change with the observation index n, because of the adaptation of the quantizer. Although important refinements can be added, as discussed below, the major factor in the adaptation of the quantizer at index n is based on the previous decision at index $(n-1)$ about whether or not an impulse was present. An impulse is judged to occur if the current data element lies outside the current confidence interval.

In order to describe the adaptation more explicitly we consider the following two cases:

Case (1): If we have decided that the previous data sample was not contaminated by an outlier then the formula for the quantizer output, an estimate of the smooth component, x_n, is

$$\tilde{x}_n = W_1(\Delta_n) + y_{n-1} \tag{6}$$

where

$$W_1(\Delta_n) = \begin{cases} \Delta_n & , \text{ if } -\tau_1 \tilde{S}_n \le \Delta_n \le \tau_1 \tilde{S}_n \\ 0 & , \text{ otherwise} \end{cases} \tag{7}$$

in which $\tilde{S}_n$ is the robust estimate of formula (3) and τ_1 is a threshold constant. The function W_1 performs the elimination or blanking of an impulse. A more analytical treatment of this blanking appears later in this Section.

Case (2): If we have decided that the previous data sample was contaminated by an outlier, (that is, blanking occurred on the previous sample), then we use zero (the mean value of x_n) as the conditional mean value of x_n and the estimate of x_n, the smooth component of the data, is

$$\tilde{x}_n = W_2(y_n) \tag{8}$$

where

$$W_2(y_n) = \begin{cases} y_n & , \text{ if } -\tau_2 \tilde{V}_n \le y_n \le \tau_2 \tilde{V}_n \\ 0 & , \text{ otherwise} \end{cases} \tag{9}$$

$\tilde{V}_n$ is the robust estimate of the standard deviation of x_n and τ_2 is the threshold constant. In our experimental results, τ_1 and τ_2 are set assuming that x_n and Δ_n are Gaussian and zero mean.

The adaptive differential quantizer adapts by estimating confidence bounds on the range of the smooth component, x_n, of the next observed data sample y_n using robust estimates of the standard deviations of x_n and its rate of change r_n based on N previously observed data samples. Of course, a more sophisticated decay of the estimated smooth component $\tilde{x}_n$ could be used during a burst of outliers, if more prior information were available.

The adaptive differential quantizer was applied to arctic acoustic data containing transients and impulses. The plots of the data before and after processing in Fig. 1 indicate that the special differential quantizer functioned well and was robust in removing the transients.

III. THEORY OF IMPULSE BLANKING WITH AND WITHOUT ADDITIONAL CORRELATED INTERFERENCE

In this section we reconsider the problem of the previous two sections using more specific assumptions about the signal and noise. This permits us to obtain some analytical results.

The realization that many important communications channels cannot be modeled as Gaussian has led to a substantial amount of research in the areas of non-Gaussian and robust detection [e.g. 23,24]. In this section we present examples of the performance of several receiver structures for the detection of a known signal in an additive impulsive environment. The noise is modeled as being nominally Gaussian with sporadic occurrences of higher variance samples which we call impulses.

The philosophy employed in the proposed receiver design is to detect the occurrences of the impulses and eliminate them from the data. Although this philosophy is at odds with researchers who attempt to accommodate bad data samples, it is felt that the lack of an accurate statistical description for the impulsive component, along with the general lack of useful information in the data containing impulses, justify its use. For an informative and interesting discussion of this issue the reader is referred to [26]. To obtain performance results, a Gaussian model was assumed for the impulsive component. Receiver operating characteristic (ROC) curves [27] (probability of detection, P_d, versus probability of false alarm, P_{fa}) are provided for both independent and correlated noise examples.

The detection problem is characterized as the binary hypothesis test

$$
\begin{aligned}
H_0 &: \; x(j) = n(j) + i(j) \\
& \qquad\qquad j = 1, 2, \cdots N \qquad (10) \\
H_1 &: \; x(j) = n(j) + i(j) + s
\end{aligned}
$$

The background noise sequence $\{n(j)\}$ is assumed to be Gaussian with mean zero, variance σ^2, and covariance matrix R. The impulsive noise term, $i(j)$, is usually zero and is occasionally a sample from an independent Gaussian source with mean zero and variance $K\sigma^2$, $K >> 1$. For sporadic impulses, we assume that $i(j)$ is non-zero ϵ of the time ($o < \epsilon < 1$, typically near zero) for a total of $N\epsilon$ impulse-corrupted samples in an observation of length N.

We first consider the clarivoyant case in which we are given knowledge of when the impulses occur. Then $\{n(j) + i(j)\}$ is a nonstationary Gaussian sequence with zero mean and covariance matrix

$$P = R + K\sigma^2 \sum_{i=1}^{N\epsilon} \underline{u}_i \underline{u}_i^T \qquad (11)$$

where $\underline{u}_i$ is an N element column vector containing a one in the sample number of the i-th impulse and zeroes elsewhere. The optimum clairvoyant test statistic for this environment is $\underline{x}^T P^{-1} \underline{s}$ which has detection performance

$$P_d = Q(Q^{-1}(P_{fa}) - \underline{s}^T P^{-1} \underline{s}) \qquad (12)$$

with $Q(\cdot)$ the Gaussian tail probability and false alarm probability denoted by P_{fa}. Although physically unrealizable, because we do not know when the impulses occur, this detector yields an upper bound of performance. For comparison, the likelihood-ratio test statistic for the background source is $\underline{x}^T R^{-1} \underline{s}$. This is a matched filter detector designed for the background noise only. With impulses present its' performance is

$$P_d = Q(Q^{-1}(P_{fa}) - \underline{s}^T R^{-1} \underline{s} / \underline{s}^T R^{-1} P R^{-1} \underline{s}) \qquad (13)$$

Define the eliminator as that suboptimum clairvoyant detector which ignores the impulsively corrupted samples. The decision, then, is based on $N(1 - \epsilon)$ samples from the background source and is independent of the variance of the impulses. In the examples below this detector demonstrates that that portion of the data corrupted by the impulses contributes little to the efficiency of the decision. Unfortunately, the eliminator is not physically realizable. Hence we consider realizable approximations to the eliminator based on the estimation of the occurrences of impulses.

A realizable approximation to the eliminator for mutually independent background noise values $\{n(j)\}$ is a blanker which ignores all samples with magnitude greater than $b\sigma$

where σ denotes the standard deviation of $n(j)$ (see the inset in Figure 2). Observations which are large in magnitude are most likely due to impulses and are removed. Unfortunately impulses which are small in magnitude are passed and large non-impulsive observations are blanked causing the performance of the blanker to fall below that of the eliminator. Due to the complexity of the density function of the resulting test statistic, asymptotic (Central Limit Theorem) results were generated for sample length $N = 50$, with 5 impulsive samples ($\epsilon = 0.1$), and blanking range $b = 4$ for impulse variance multipliers of $K = 25$ and 100. These ROC curves appear in Figure 2 along with the exact curves for the optimum clairvoyant (formula 12), clairvoyant eliminator, and matched filter (formula 13) detectors.

To approximate the performance of the clairvoyant eliminator in correlated noise we estimate the occurrences of impulses using the system inset in Figure 3 which contains a blanker imbedded within a whitening loop for the background noise $n(j)$. This is closely related to the adaptive differential quantizer of the previous section. Assuming that the blanker has eliminated previous impulses, the whitener outputs the innovations sequence for the background source plus any current impulse in the data. If the signal is weak, it has little effect on this output. As in the independent case, the blanker estimates the occurrence of an impulse by a magnitude test on the output of the whitener. This approach was proposed by Tufts and a preliminary study was made by Kirsteins and Tufts [6]. For robustness, they used an integrator instead of a perfect whitener and some concepts from adaptive differential quantization. These modifications are useful when the signal is not weak and the correlation of the background noise is not well known or is changing. Some of these practical modifications were discussed in the previous section.

Computation of detector performance for a correlated environment depends directly upon the environment parameters. As an example, we consider a ten-sample observation from a first order autoregressive source $\{n(j)\}$ with coefficient $a = 0.7$. Typically, if a sample is blanked, the zero fed into the whitening loop causes a decrease in the whitener's performance. We remove that problem from this example by allowing the impulsive source to contaminate only the last data point. Further, we assume that none of the first 9 data points are blanked. For a blanker limit of 4σ (equivalent to approximately 8 times the standard deviation of the innovations sequence) this assumption is true with probability $> 1 - 10^{-10}$. The blanker, then, determines the performance through its effects on the last sample. Plots of the ROC for this detector, the matched filter, and the clairvoyant detectors appear in Figure 3 for $K = 25$ and 100.

The clairvoyant detectors provide upper bounds, though unrealizable, to detector performance while the matched filter provides a realizable lower bound. Figures 2 and 3 demonstrate that the technique of estimating when impulses occur and eliminating those data points yields very good performance, close to the unrealizable performance upper bound. Optimal use of all of the data in the test statistic as in [28] should provide a gain in performance. However, the magnitude for any increase is limited by the clairvoyant detector bound and is, in practice, often cancelled by our incomplete knowledge of the impulsive source's statistics. In this sense the estimation-elimination technique displays robustness to the impulsive environment.

IV. EFFICIENT REMOVAL OF THE NARROW BAND AND NARROW ANGLE COMPONENTS BY SIMPLIFIED COMPUTATION OF PRINCIPAL EIGENVECTORS AND THEIR EIGENVALUES

Having discussed the removal of impulse noise, which is the first step in our proposed detection procedure, we now return to the procedures for removing the coherent or signal-like portions of the interference. These components are characterized by spatial or temporal covariance matrices which are approximately of low rank, that is a rank much less than the size of the covariance matrix. Such components can be removed by (a) projecting successive data vectors on the estimated eigenvectors associated with the larger eigenvalues and (b) subtracting these components from the data [8,13,21,22]. If, in addition, the components are exponential in time (damped sinusoids) or space (plane waves seen by line arrays) then our improved versions of Prony's method can be used [9,10,11,12].

Here we suggest different approaches to achieving high-resolution performance without the computational cost of actually computing the singular value decomposition SVD or computing eigenvectors and eigenvalues [17]. The ideas are based on the power method and on a method of Lanczos [15,16]. We are motivated by VLSI realization of high-resolution signal processing [29].

Prony's Method and the Weigthed Power Sums: A Derivation of Lanczos Method

Let us assume that we start with a given matrix A for which we want to compute the principal eigenvectors and eigenvalues. Let us also define the eigenvectors and eigenvalues associated with the conjugate-symmetric square matrix A ($dim\ A = n$).

$$A\underline{u}_i = \lambda_i \underline{u}_i,\ i = 1,2,...,n \tag{14}$$

where

$$\begin{aligned} \underline{u}_i^* \cdot \underline{u}_j &= 0,\ i \neq j \\ \underline{u}_i^* \cdot \underline{u}_j &= 1,\ i = j \end{aligned} \tag{15}$$

The characterization polynomial associated with the matrix A is given by

$$det(A - \lambda I) = 0 \tag{16}$$

Expanding the determinant we have the polynomial equation

$$\lambda^n + p_{n-1}\lambda^{n-1} + \cdots + p_o = 0 \tag{17}$$

and the roots of this polynomial will give us the eigenvalues λ_i of the matrix. We briefly summarize the procedure for obtaining the eigenvalues λ_i for Hermitian Matrices based on the Lanczos "power sums" as presented in [15]. We shall show that the eigenvalues can then be obtained from the power sums by Prony's method [14].

Let us select a starting vector $\underline{b}_o$. There is an implicit assumption in the Lanczos procedures that the starting vector $\underline{b}_o$ has a non-zero projection on some eigenvector of the matrix A corresponding to each eigenvalue that we want to compute. We usually obtain the starting vector from a preliminary Fourier analysis of the data.

We then analyze the vector $\underline{b}_o$ in the reference system of the base vectors $\{\underline{u}_i\}$, which are the set of orthonormal eigenvectors of the matrix A:

$$\underline{b}_o = \tau_1 \underline{u}_1 + \tau_2 \underline{u}_2 + \cdots + \tau_n \underline{u}_n \tag{18}$$

where

$$\tau_i = \underline{u}_i^* \underline{b}_o \tag{19}$$

Hence, using equation (14), successive vectors formed by premultiplications of $\underline{b}_o$ by powers of the matrix A can be represented as follows:

$$\begin{aligned} \underline{b}_1 = A\underline{b}_o &= \tau_1\lambda_1\underline{u}_1 + \tau_2\lambda_2\underline{u}_2 + \cdots + \tau_n\lambda_n\underline{u}_n \\ &\cdot \\ &\cdot \\ &\cdot \\ \underline{b}_{k+1} = A^k\underline{b}_o = A\underline{b}_k &= \tau_1\lambda_1^k\underline{u}_1 + \tau_2\lambda_2^k\underline{u}_2 + \cdots \\ &+ \tau_n\lambda_n^k\underline{u}_n \end{aligned} \tag{20}$$

Let us form the set of basic scalars:

$$c_{i+k} = \underline{b}_i^* \underline{b}_k = \underline{b}_k^* \underline{b}_i \tag{21}$$

Then we shall have:

$$c_k = |\tau_1|^2\lambda_1^k + |\tau_2|^2\lambda_2^k + \cdots + |\tau_n|^2\lambda_n^k = \underline{b}_o^* A^k \underline{b}_o \tag{22}$$

which were called by Lanczos the "weighted power sums" [15]. The problem of obtaining λ_i's from the c_i's is the "problem of weighted moments" [15]. That is it is the problem of Prony and the old and modern versions of Prony's method can be used to estimate the λ_i's from matrix C [12,14].

Thus in the absence of noise we know that using the weighted power sums c_k of (21) as input data for the improved Prony method [11,12] and finding the roots of the resulting polynomial will provide us with accurate estimates of the true eigenvalues λ_i of the covariance matrix A. We have found experimentally that, in the presence of noise the method still provides us with accurate solutions for our problem of estimating the frequencies of one or two sinusoids, provided the trial vector $\underline{b}_o$ is biased in favor of the signal subspace of one or two largest eigenvectors. The actual eigenvectors can be rotated because of the scale factor of $\underline{b}_o$, but the subspace is accurately defined. To obtain the eigenvectors we use the approximate identity.

$$\prod_{i=1}^{n} (A - \lambda_i I) A\underline{b}_0 = 0 \tag{23}$$

This suggests a method for estimating the eigenvectors $\underline{u}_1$, $\underline{u}_2$. Using equation (14), we obtain an approximation to the eigenvector $\underline{u}_1$ (associated with λ_1) by the following formula:

$$\tilde{u}_1 = (A - \lambda_2 I)Ab_o = A^2 b_o - \lambda_2 A b_o = b_2 - \lambda_2 b_1 \tag{24}$$

Similarly for the eigenvector u_2 (associated with λ_2):

$$\tilde{u}_2 = (A - \lambda_1 I)Ab_o = A^2 b_o - \lambda_1 A b_o = b_2 - \lambda_1 b_1 \tag{25}$$

For the one sinusoid case, the rank of A is approximately one and hence equation (23) becomes:

$$(A - \lambda_1 I)\, Ab_o = 0 \tag{26}$$

Hence for the problem of one or two sinusoids, by solving only a quadratic equation we can obtain the eigenvectors associated with the one or two principal eigenvalues exactly.

The Power Method and VLSI Realization

Suppose A is a Hermitian $(n \times n)$ matrix. The SVD theorem states that A can be written as:

$$A = U \cdot S \cdot U^T \tag{27}$$

where U is a unitary matrix and S is a matrix consisting of only real diagonal elements [17].

The power method computes the dominating singular vectors one at a time and is based on solving the equation

$$su = Au \tag{28}$$

for the singular vector u and the singular value s. The power method uses an iterative scheme to solve (28). We instead suggest a two-step solution using an appropriate starting vector b_o:

$$\tilde{u}_1 = A\ b_o / \| A\ b_o \| \tag{29}$$

The singular value is chosen to be:

$$s_1 = \| A\ b_o \| \tag{30}$$

In order to obtain the next singular vector, the estimated singular plane $(u_1 u_1^T)$ is removed from A using the following deflation procedure [17]:

$$A' = A - s_1 u_1 u_1^T \tag{31}$$

and the procedure is repeated with matrix A' to yield s_2, u_2.

The selection of b_o is very important and the Fourier vector provides a very good estimate. This preprocessing step can be implemented in VLSI very efficiently using summation-by-parts [18] or the Fast Hartley Transform [19,20] methods.

REFERENCES

1. Wigner, Eugene, Nobel Prize address, 1963.

2. Liu, S.C. and Nolte, W.N., "Performance Evaluation of Array Processors for Detecting Gaussian Acoustic Signals," *IEEE Trans. ASSP,* Vol. ASSP-28, No. 3, June 1980.

3. Claus, A.J., Kadota, T.T., and Romain, D.M., "Efficient Approximation for a Family of Noises for Application in Adaptive Spatial Processing for Signal Detection," *IEEE Trans. on Information Theory,* Vol. IT-26, pp. 588-595, September 1980.

4. Martin, R.D. and Thomson, D.J., "Robust-Resistant Spectrum Estimation," *Proc. of IEEE,* Vol. 70, No. 9, September 1982.

5. Veitch, J.G. and Wilks, A.R., "A Characterization of Arctic Undersea Noise," Office of Naval Research, Princeton University, Report No. 12.

6. Kirsteins, I. and Tufts, D.W., "Methods of Computer-Aided Analysis of Non-Gaussian Noise and Application to Robust Adaptive Detection," *Tech. Report - Part II,* Dept. of Electrical Engineering, University of Rhode Island, October 1984.

7. Swaszek, P.F., Tufts, D.W. and Efron, A.J., "Detection in Impulsive Environments," *Letters, Proceedings of the IEEE,* Vol. 73, No. 12, December 1985.

8. Tufts, D.W. Kumaresan, R., and Kirsteins, I., "Data Adaptive Signal Estimation by Singular Value Decomposition of a Data Matrix," *Proceedings of the IEEE,* Vol. 70, No. 6, June 1982.

9. Tufts, D.W. and Kumaresan, R., "Estimation of Frequencies of Multiple Sinusoids: Making Linear Prediction Perform Like Maximum Likelihood," *Proceedings of the IEEE,* Vol. 70, No. 9, September 1982.

10. Kumaresan, R. and Tufts, D.W., "A Two Dimensional Technique for Frequency Wavenumber Estimation," *Proceedings of the IEEE,* Vol. 69, No. 11, November 1981.

11. Kumaresan, R. and Tufts, D.W., "Estimating the Parameters of Exponentially Damped Sinusoids and Pole-Zero Modeling in Noise," *IEEE Trans. on Acoustics, Speech, and Signal Processing,* Vol. ASSP-30, No. 6, December 1982.

12. Kumaresan, R., Tufts, D.W. and Scharf, L.L., "A Prony Method for Noisy Data: Choosing the Signal Components and Selecting the Order in Exponential Signal Models," *Proceedings of the IEEE,* Vol. 72, No. 2, February 1984.

13. Tufts, D.W., Kirsteins, I., and Kumaresan, R., "Data-Adaptive Detection of a Weak Signal," *IEEE Trans. on Aerospace and Electronic Systems,* Vol. AES-19, No. 2,

March 1983.

14. Hildebrand, F.B., Introduction to Numerical Analysis, McGraw Hill, New York, 1956.

15. Lanczos, C., Applied Analysis, Prentice-Hall, Inc., 1956.

16. Parlett, B., The Symmetric Eigenvalue Problem, Prentice Hall, 1980.

17. Golub, G.H. Van Loan, C.F., Matrix Computations, John Hopkins University Press, 1983.

18. Boudreaux, G.F., Parks, T.W., "Discrete Fourier Transform Using Summation by Parts," submitted to *IEEE Trans. ASSP.*

19. Bracewell, R.N., "The Fast Hartley Transform," *Proc. IEEE,* Vol. 72, No. 8, August 1984.

20. Kumaresan, R. and Gupta, P.K., "A Real-Arithmetic Prime Factor Fourier Transform Algorithm and Its Implementation," submitted to *IEEE Trans. ASSP.*

21. Kirsteins, I.P. and Tufts, D.W., "On the Probability Density of Signal-to-Noise Ratio in an Improved Adaptive Detector," presented at *ICASSP 85,* Tampa, Florida, March 1984.

22. Kirsteins, I.P. and Tufts, D.W., "On the Probability Density of Signal-to-Noise Ratio in an Improved Adaptive Detector," *Tech. Report - Part I,* Dept. of Electrical Engineering, University of Rhode Island, October 1984.

23. Miller, J.H. and Thomas, J.B., "Detectors for Discrete-Time Signals in Non-Gaussian Noise," *IEEE Trans. Inform. Theory,* Vol. IT-18, pp. 241-250, March 1972.

24. Martin, R.D. and Schwartz, S.C., "Robust Detection of a Known Signal in Nearly Gaussian Noise," *IEEE Trans. Inform. Theory,* Vol. IT-17, pp. 50-56, January 1971.

25. Dwyer, R.F., "A Technique for Improving Detection and Estimation of Signals Contaminated by Under Ice Noise," *NUSC Tech. Doc. #6717,* July 1982.

26. Beckman, R.J. and Cook, R.D., "Outlier..........s," *Technometrics,* Vol. 25, pp. 119-163, February 1983.

27. Van Trees, H.L., Detection, Estimation, and Modulation Theory, Pt. 1, New York, Wiley, 1968.

28. Miller, J.H. and Thomas, J.B., "The Detection of Signals in Impulsive Noise Modeled as a Mixture Process," *IEEE Trans. Commun., Vol. COM-24, pp. 559-563, May 1976.*

29. D.W. Tufts and C.D. Melissinos, "Simple, Effective Computation of Principal Eigenvectors and their Eigenvalues and Application to High-Resolution Estimation of Frequencies," resubmitted to *IEEE Trans. ASSP,* August 1985.

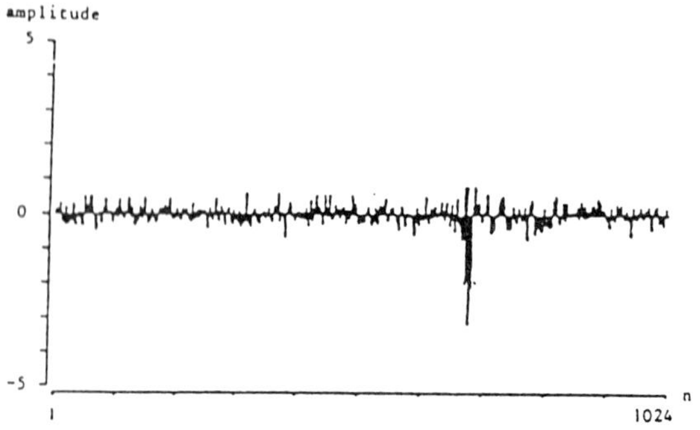

a) Record #1362: Coefficient of kurtosis = 28.2

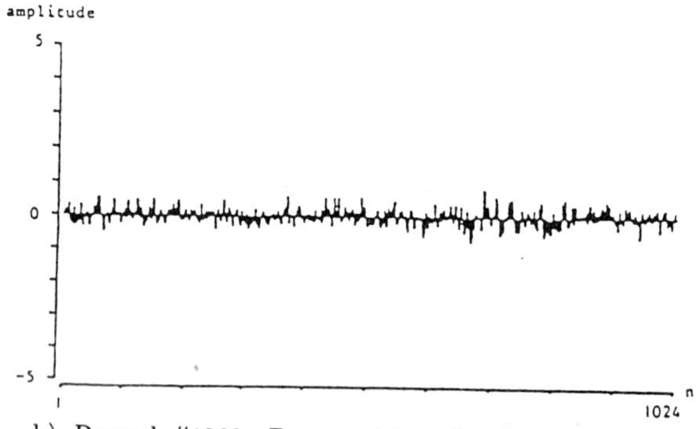

b) Record #1362: Processed by adaptive differential quantization where T_1=3.76, T_2=2.51, and N=101 coefficient of kurtosis = 3.19

Figure 1 Waveform plots of records containing impulsive bursts before and after processing by adaptive differential quantization.

NOTE: The coefficient of kurtosis is defined as

$$c_k = \frac{m_4}{(m_2)^2}$$

where

$$m_1 = \frac{1}{1024}\sum_{n=1}^{1024} x_n \qquad m_2 = \frac{1}{1024}\sum_{n=1}^{1024}(x_n - m_1)^2$$

$$m_4 = \frac{1}{1024}\sum_{n=1}^{1024}(x_n - m_1)^4$$

and x_n, vor n=1,2...1024 are samples of the particular noise record.

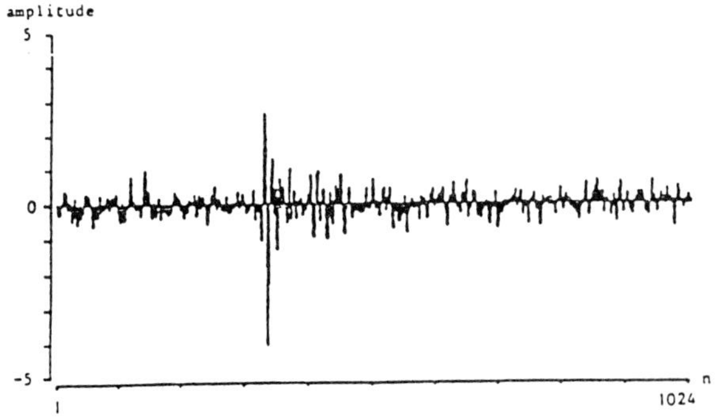

c) Record #2066: Coefficient of kurtosis = 29.52

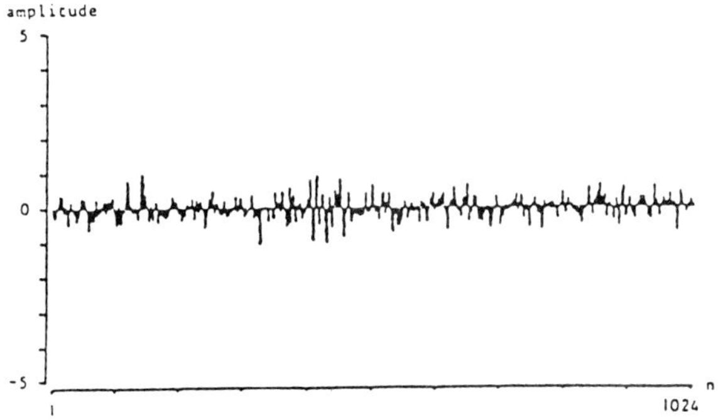

d) Record #2066: Processed by adaptive differential quantization where T_1=3.76, T_2=2.51, and N=101 coefficient of kurtosis = 3.91

Figure 1 Waveform plots of records containing impulsive bursts before and after processing by adaptive differential quantization.

NOTE: The coefficient of kurtosis is defined as

$$c_k = \frac{m_4}{(m_2) sup\, 2}$$

where

$$M_1 = \frac{1}{1024} \sum_{n=1}^{1024} x_n \qquad m_2 = \frac{1}{1024} \sum_{n=1}^{1024} (x_n - m_1)^2$$

$$M_4 = \frac{1}{1024} \sum_{n=1}^{1024} (x_n - m_1)^4$$

and x_n, for n=1,2...1024 are samples of the particular noise record.

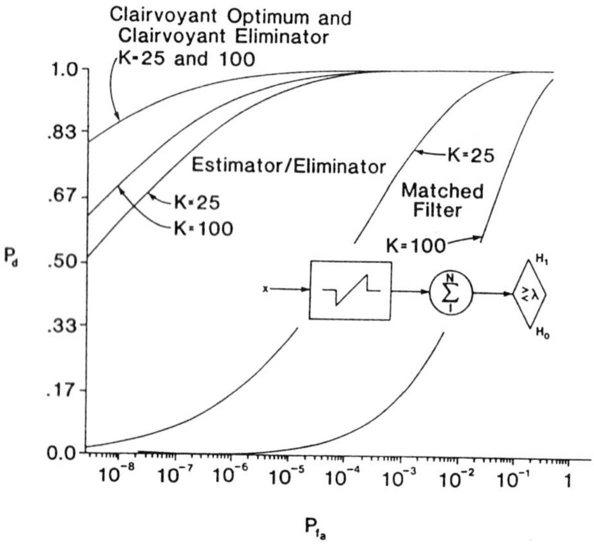

Figure 2 Independent noise detector and ROC curves, $s = 1$, $\sigma^2 = 1$

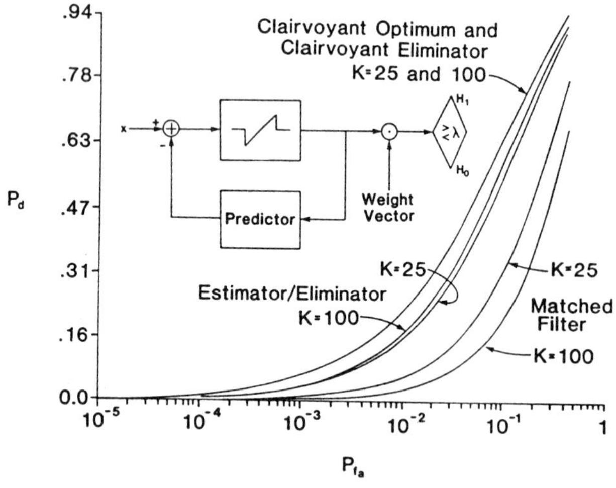

Figure 3 Correlated noise detector and ROC curves, $s = 1$, $\sigma^2 = 1$

ON NONGAUSSIAN SIGNAL DETECTION AND CHANNEL CAPACITY

Charles R. Baker *
Department of Statistics
University of North Carolina
Chapel Hill, NC 27514

I. INTRODUCTION

This paper contains a discussion of some recent research in signal detection and communications. No proofs are included; the emphasis is on motivation and results. Some precise definitions and results are contained in the Appendix. For a more detailed discussion and additional results, reference is made to [8,9,10] for the signal detection problems, and to [7,8] for the channel capacity problems. The present paper is couched in terms of problems in underwater acoustics; as will be seen, the models and results are of general applicability.

II. SIGNAL DETECTION

Two problems will be discussed. The first, for which the most complete results were obtained, was that of detecting nonGaussian signals in Gaussian noise. The second, motivated by examination of noise properties for actual sonar data, was that of detection in a class of nonGaussian processes, which can be regarded as mixtures of Gaussian processes.

Both of these problems are important in sonar applications. The detection of nonGaussian signals in Gaussian noise can be regarded as the canonical problem for active detection in reverberation-limited noise, especially volume reverberation. The noise in such situations can frequently be regarded as arising from reflections by many small scatterers, which can be reasonably assumed to have statistically-independent behavior. The central limit theorem then gives a Gaussian process. The signal process, however, will frequently be dominated by reflections from a few large scatterers, such as the sonar dome. These scatterers each give rise to a nonGaussian random process, which are summed at the receiver to give a nonGaussian process.

Other applications may also involve detection of nonGaussian signals in Gaussian noise. For example, an emerging passive sonar detection problem is that of detecting very quiet submarines, emanating primarily broadband signals. These signals may prove to be nonGaussian and one will frequently be faced with detecting them in a Gaussian noise background.

The importance of the problem of detecting signals in nonGaussian noise of a "spherically-invariant" (Gaussian mixture) type has become apparent by examining the results of data analysis on acoustic recordings obtained from both under-ice and shallow-water environments. These noise recordings have exhibited data whose

* Research supported by ONR Contracts N00014-75-C-0491, N00014-81-K-0373, and N00014-84-C-0212.

univariate distribution properties appear similar to those of Gaussian random variables (symmetric, unimodal, smooth). However, when compared with zero-mean Gaussian random variables having the same variance, the data often exhibits heavy tails and/or high kurtosis. These features, at least in the univariate case, are very appropriate to a Gaussian mixture model for the noise. If the multivariate data has a Gaussian mixture distribution, then the modeling problem (in the context of modeling nonGaussian processes) is greatly simplified, as will be discussed below.

III. DETECTION OF NONGAUSSIAN SIGNALS IN GAUSSIAN NOISE

Our objective here was to give a complete solution of the detection problem. The actual data processes appear as functions of continuous time. The desirable results then include the following:

(1) Characterization of signal-plus-noise processes for which the detection problem is well defined. By this, we mean a mathematical model which does not promise perfect (singular) detection. Such singular models are not considered to be realistic.

(2) For well-defined problems, determination of the likelihood ratio for the continuous-time problem.

(3) Approximation of the continuous-time likelihood ratio by a discrete-time form, preferably in recursive or near-recursive form.

(4) Specification of procedures for estimating parameters appearing in the approximation to the likelihood ratio.

(5) Performance evaluation of the approximation to the likelihood ratio.

Of course, there are other desirable results, such as development of robust approximations to the likelihood ratio, and approximations which do not require a full description of data parameters. However, the results listed in (1)-(5) above are already very ambitious, and obtaining them would be a significant step in any complete solution to the problem.

Considerable work has been done on detection of Gaussian signals in Gaussian noise; see for example some of the references given in [5]. However, previous work on detection of nonGaussian signals in Gaussian noise has been subject to one or more of the following limitations:

(a) The noise is assumed to be the Wiener process; see, e.g., [15]. The paths of the Wiener process are far too irregular to reasonably model sonar noise. Other Wiener process properties, such as independent increments, Markov, etc., are not typically satisfied. Moreover, determination of likelihood ratio parameters is left as an open problem.

(b) Detection is based on second-moment criteria, such as the deflection criterion [1], [2].

(c) Signal and noise are taken to be independent [3], and expressions for the likelihood ratios are not obtained.

We have obtained a reasonably complete solution to problems (1)-(3) above. The solutions to (4) and (5) are presently being computationally investigated. Here we give a rough summary of the results. More precise statements require a substantial mathematical machinery; reference is made to [9] for the complete and final statements; a more accessible treatment, with more reference to sonar applications, is contained in [10]. Partial results are contained in the Appendix.

The data is assumed to be observed over a finite interval, which we take as [0,1]. The noise is Gaussian, mean-square continuous, zero-mean, and is assumed to vanish (almost surely) at t=0.

To solve problem (1) mentioned above, one would wish to consider a general nonGaussian process (Y_t), and determine necessary and sufficient conditions for the detection problem to be well-defined (non-singular). Such conditions are given in Theorem 1 and Theorem 2 of the Appendix. Roughly, they require that the process (Y_t) have a signal-plus-noise representation $Y_t = S_t + N_t$, where the sample paths of (S_t) belong almost surely to the reproducing kernel Hilbert space of (N_t). This condition means that the covariance function (resp., sample paths) of the signal process must be much smoother than the covariance (resp., sample paths) of the noise. The process (S_t) must also satisfy certain measurability conditions with respect to (Y_t) and (N_t). See the Appendix for the precise statements. We remark that the necessary conditions and the sufficient conditions are not identical, although they are very close.

Problem (2) mentioned above is that of determining the continuous-time likelihood ratio. A general solution has been obtained, and is given in [9]. Actually, two solutions are given there. One views the observations as being simply real-valued functions; the other treats them as being elements of $L_2[0,1]$. The latter is summarized in the Appendix. Here we shall give the finite-sample discrete-time approximation to the likelihood ratio on $L_2[0,1]$.

First, the noise is a Gaussian vector $\underline{N}$ having covariance matrix $\underline{R}$. We can represent $\underline{R}$ by $\underline{R} = \tau^2 \underline{E}\underline{E}^*$ where E is a lower-triangular matrix, and τ is the sampling interval. We can thus consider the noise process to be a sampled version of $N_t = \int_0^t F(t,s)dW_t$, where (W_s) is the standard Wiener process. According to the results of [9], the signal-plus-noise process will be of the form $S_t + N_t = \int_0^t F(t,s)dZ_s$, where (Z_t) is here taken to be a diffusion with memoryless drift function σ: $Z_t = \int_0^t \sigma(Z_s)ds + W_t$. The resulting discrete-time approximation to the log-likelihood ratio is then

$$\Lambda^{n+1}(X^{n+1}) = \Lambda^n(X^n) + \tau^{1/2}\sigma[\tau^{1/2}(\underline{L}\underline{E}_n^{-1}X^n)_n](E_{n+1}^{-1}\, X^{n+1})_{n+1}$$

$$- \frac{1}{2}\tau\sigma^2[\tau^{1/2}(\underline{L}\underline{E}_n^{-1}X^n)_n], \quad n \geq 1;$$

$$\Lambda^1(X^1) = 0$$

where X^n denotes the observation vector obtained from the first n samples;

$\tau^2 E_n E_n^*$ is the noise covariance matrix for the first n sample times;

L is the summation matrix: $(LX^n)_i = \sum_{j=1}^{i} X_j^n$.

This formulation of the log-likelihood ratio is partially recursive. Note that

$$(LE_n^{-1} X^n)_n = (LE_{n-1}^{-1} X^{n-1})_{n-1} + (E_n^{-1} X^n)_n,$$

and that the operation $(E_n^{-1} X^n)_n$ is just a cross-correlation of the data vector with the nth row of E_n^{-1}.

There are three basic considerations in evaluating the usefulness of the above log-likelihood ratio. One is the validity of the approximation assumption; a second is the development of procedures for estimating the parameters of the likelihood ratio; finally, one is interested in whether or not the discrete-time approximation is in fact a likelihood ratio when our assumptions are satisfied. We discuss these three points below.

(i) Any Gaussian vector can be obtained by passing white Gaussian noise through an appropriate lower-triangular matrix. Thus, the noise model is reasonable for the discrete-time problem, and one can justify the use of multiplicity M=1 from this and from other mathematical considerations. The fact that (Z_t) is a process of diffusion type then follows from well-known results [15]; to assume further that it is of diffusion type with respect to (W_t), one reasons that the difficult detection problems are of most interest; such problems are those in which the N and $S+N$ processes have very similar properties. Since (N_t) is modeled as a time-varying linear operation on the diffusion (W_t), it seems reasonable to model (Y_t) as that same time-varying linear operation on a process that is of diffusion type with respect to (W_t). More detailed physical interpretations of the assumptions can be given for applications in sonar. However, a basic reason for making the assumptions is that they permit one to implement an approximation to the likelihood ratio without detailed knowledge of the data probability distributions. The validity of these assumptions and the effectiveness of the finite-sample discrete-time approximations can be judged in each application by the performance of the detection algorithm.

(ii) The implementation of the sequence of test statistics (Λ^k) given above, for $k \leq n$, requires knowledge of only two parameters: the lower-triangular matrix E_n such that $\tau^2 E_n E_n^*$ is the $n \times n$ noise covariance matrix, and the drift function σ. Typically (in sonar) these quantities will need to be estimated from experimental data. We give a procedure for doing this, supposing that one has an ensemble of independent sample vectors from the noise process, of sufficient size to give a good estimate of the covariance matrix, and that one or more sample vectors from the signal-plus-noise process is available.

First, the noise vector is written as $\underline{N} = \underline{F} \, \underline{\Delta W}$, where ΔW is the vector with j^{th} component $(W[j\tau] - W[(j-1)\tau])$ for $j \geq 2$, and first component $W(\tau)$. If $i\tau = t_i$ and the ij element of $\underline{F}$ is $F(i,j)$ for all ij, then $N_i \cong N(i\tau) = N(t_i)$ for large i and small τ. The representation for $\underline{N}$ gives noise covariance matrix $R_N = \tau^2 FF^*$. Consistent with this representation and the results of [9], the $S+N$ vector is written as $\underline{Y} = \underline{F} \Delta \underline{Z}$, where ΔZ is the vector with j^{th} component $(Z[j\tau] - Z[(j-1)\tau])$ for $j \geq 2$, and first component $Z(\tau)$. Thus, given an ensemble of sample noise vectors, one treats the resulting estimate of the noise covariance matrix as R_N, and obtains the factorization $R_N = \tau^2 FF^*$. Then, given a $\underline{Y}$ sample vector, $\underline{Z}$ is estimated by

$\underline{\Delta Z} = \underline{E}^{-1}\underline{Y}$ and $(\underline{\Delta Z})_1 = Z(\tau)$. Given $\underline{Z}$, our assumptions yield

$$Z_i = Z(i\tau) = \int_0^{i\tau} \sigma(Z_s)ds + W(i\tau)$$

Various methods can then be used to estimate the unknown function σ. A general maximum-likelihood estimate is given in [12], which is now being computationally investigated. The generality of this procedure leads to computational difficulties, so that it may be necessary to assume a specific form for σ, such as a low-order polynomial with unknown coefficients.

(iii) Under our assumptions, Λ^n can be considered a "good" approximation to a likelihood ratio test statistic if τ is "small" and n is "large." These are not exact statements; at present, we have no bounds on performance. Any such bounds would involve $\underline{R}_N$, τ, σ, and n. However, if $\underline{Y}$ is Gaussian, a precise statement can be made. Suppose that $\underline{N} = \underline{E}\underline{\Delta W}$ and $\underline{Y} = \underline{E}\underline{\Delta Z}$ as above, and that (a_k) and (b_k) are two sequences of real numbers such that

$$Z^k = \sum_{j=1}^{k-1} [a_j Z_j + b_j] + W_k \qquad 2 \leq k \leq n,$$

$$Z_1 = W_1.$$

In this case, it can be shown that exp (Λ^n) is a monotone function of $dP_{\underline{Y}}/dP_{\underline{N}}$, thus a likelihood ratio test statistic. We conjecture that this also holds when $\underline{Y}$ is not Gaussian:

$$N = \underline{E}\underline{\Delta W}, \underline{Y} = \underline{E}\underline{\Delta Z}, \text{ and}$$

$$Z_k = \sum_{j=1}^{k-1} \sigma_j(Z_j) + W_k \qquad 1 \leq k \leq n,$$

with σ non-affine.

The above results give a solution to two problems of much interest in the theory of stochastic processes: determining conditions for a discrimination problem to be non-singular, and determining the likelihood ratio, when one of the two processes is Gaussian. The scope of these problems can be appreciated by reviewing some of the references cited in [5]. The above approximation to the log-likelihood ratio gives some hope of obtaining useful new detection algorithms for some important sonar detection problems. The eventual utility, however, will be apparent only after a great deal of further work is done, especially computational work involving experimental data.

IV. DETECTION IN NONGAUSSIAN NOISE

Examination of data properties by several investigators has indicated that sonar data may be spherically-invariant (a Gaussian mixture) in several important applications. One such application is in under-ice operations. Analysis of such data has shown that the univariate data typically has high kurtosis and heavy tails as compared to Gaussian data of the same variance [11].

Another environmental situation which apparently gives rise to univariate spherically-invariant noise is that of near-shore operations in warm climes (e.g., the Gulf of Mexico). The nonGaussian noise in this case is attributed to snapping shrimp and results in very high kurtosis as compared to Gaussian data [16].

These observed data properties motivated us to consider the problem of detection in spherically-invariant noise. Such a noise process (N_t) can be represented as $N_t = AG_t$, where (G_t) is a zero-mean Gaussian process with covariance function R, and A is a random variable independent of (G_t). Such processes are also said to be Gaussian-mixture processes.

Of course, the property of being univariate spherically-invariant does not imply that a process will be spherically-invariant in the multivariate case, as consideration of the case $A = 1$ will show. However, if the above representation of the noise is reasonable, then the problem of characterizing the probability distributions for nonGaussian noise is reduced to that of determining the covariance function of (G_t) and the probability distribution function of the random variable A. Without loss of generality, one can assume that the second moment $EA^2 = 1$, so that the covariance of (G_t) is the same as that of (N_t). As this function can be estimated, the major problem is that of determining the distribution function of A. We are presently carrying out computational work on this problem, using maximum-likelihood estimation.

The significance of this model, if accurate, is that it would permit one to describe all the joint distributions of the data through knowledge of the covariance (as in the Gaussian case) and of the distribution function for a single random variable. It can thus be viewed as a first step away from the Gaussian noise hypothesis which does not require that one take independent samples.

We are interested here in obtaining the same results (1)-(5) discussed above for the Gaussian noise case. So far, partial results have been obtained for (1), (2) and (3). We have found [8,10] that the sufficient conditions for a well-defined (non-singular) detection problem are the same as those obtained for detection in Gaussian noise (which are very close to being necessary). An expression for the continuous-time likelihood ratio has also been found. For detection of a known signal, an expression for the likelihood ratio has been obtained which does not require knowledge of the distribution of the random variable A. This formulation is readily approximated in the discrete-time finite-sample size case [10]. The remaining problems in obtaining (4)-(5) above have yet to be seriously investigated.

V. MUTUAL INFORMATION AND CHANNEL CAPACITY

Work to be discussed here was again for two types of noise processes: one where the channel noise is Gaussian, the other where it is spherically-invariant.

The capacity of a communication channel is here taken to be its information capacity:

$$C = \sup_Q I(m, Y)$$

where m is the message process, $A(m) = S$ is the transmitted signal, N is the channel noise, $Y = A(m) + N$ is the received process, and $I(u, v)$ is the mutual information between stochastic processes u and v (as defined in [4]). The constraint class Q contains all admissible message processes m and coding functions A. It is usually chosen from considerations involving average power, so typically involves a relation between the signal process and the noise covariance. In the case of stationary signal and noise processes, with spectral density functions Φ_S and Φ_N, an appropriate constraint under some assumptions is

$$\int_{-\infty}^{\infty} \frac{\Phi_S}{\Phi_N}(\lambda) d\lambda \leq P.$$

This can be related to the reproducing kernel Hilbert space of the noise process, and a related general constraint is $E\|A(m)\|_N^2 \leq P$, where $\|u\|_N$ is the reproducing kernel Hilbert space (for N) norm of the function $u(t)$.

If one considers this in physical terms for the frequency domain, such a constraint places a limitation on the expected value of the <u>integrated ratio</u> of signal energy to noise energy. The constraint thus involves bandwidth as well as total signal energy. In non-white noise, this is obviously more realistic than a limitation on total signal energy alone.

VI. MISMATCHED CHANNELS

With the type of constraint discussed above, a complete solution to the channel capacity problem for Gaussian channels without feedback is given in [4]. However, this approach will not be valid when the covariance of the channel noise (N_t) is unknown. This can occur from natural causes, as with insufficient knowledge of the environment. It can also occur because of jamming in the channel. In the latter case, it is well-known that if the channel noise has a given covariance, then channel capacity is minimized when the noise is Gaussian. Thus, a jammer seeking to minimize capacity of a channel with ambient Gaussian noise would choose to add Gaussian noise, and the channel capacity would then be determined by the relation between the actual channel noise (including the jammer's contribution) and the noise covariance assumed by the user of the channel. Of course, less obvious questions also arise. Channel capacity is only the starting point in analyzing such situations.

These considerations have motivated us to introduce the notion of "mismatched" channels, wherein the constraint on transmitted signals is taken with respect to a covariance which is different from that of the channel noise.

An analysis of this problem is contained in [6] for a large class of Gaussian noise processes. Additional results are forthcoming [7]. Striking differences appear between the results for the mismatched channel and those for the matched channel (when the channel noise is also the constraint noise). For example, in the matched continuous-time channel with the above generalized power constraint $(E\|A(m)\|_N^2 \leq P)$, the capacity of the channel is equal to $P/2$ and cannot be actually attained. In the mismatched channel, the capacity can be either greater or smaller than the capacity for the matched channel and it can be attained in some situations. The value of the capacity depends on the relation between the two covariances. We give one result from [6].

Let the constraint covariance operator in $L_2[0,T]$ be denoted by R_W (consider W as the noise assumed by the channel user) and let R_N be the covariance operator for the channel noise process N. Suppose that $R_N = R_W^{1/2}(I+S)R_W^{1/2}$ where I is the identity in $L_2[0,1]$ and S is a compact operator. This relation will be satisfied, for example, when W and N are two Gaussian processes for which the discrimination (detection) problem is well-defined. Let

$$C_W(P) = \sup_Q I[A(m), Y]$$

when Q contains all coding operations A and stochastic processes m (including nonGaussian processes) on [0,T] such that $E\|A(m)\|_W^2 \leq P$. Finally, let $\{\lambda_n, n \geq 1\}$ denote the strictly negative eigenvalues of the operator S defined above. Of course, this set may be empty, as when N can be written as $N = W+V$, with V independent of W. Let $\{e_n, n \geq 1\}$ be associated o.n. - eigenvectors. Then [6]:

(a) If $\{\lambda_n, n \geq 1\}$ is not empty and $\sum_n |\lambda_n| \leq P$, then $C_W(P) = \frac{1}{2}\sum_n \log[(1+\lambda_n)^{-1}] + \frac{1}{2}[P + \sum_m \lambda_m]$.

(b) If $\{\lambda_n, n \geq 1\}$ is not empty, and $\sum_n |\lambda_n| > P$, then there exists a largest integer K such that $\sum_1^K \lambda_i + P \geq K\lambda_K$, and

$$C_W(P) = \frac{1}{2}\sum_{n=1}^{K} \log\left[\frac{\sum_1^K \lambda_i + P + K}{K(1+\lambda_n)}\right]$$

(c) If $\{\lambda_n, N \geq 1\}$ is empty, $C_W(P) = P/2$.

(d) In (a) and (b), the capacity is strictly greater than when $R_N = R_W$; in (c) these capacities are equal.

(e) In (a), the capacity can be attained if and only if $\sum_n |\lambda_n| = P$. It is then attained by a Gaussian signal with covariance operator $R = \sum_{i=1}^K \beta_i u_i \otimes u_i$, where $u_n = Ue_n$, U unitary, and
$\beta_n = -\lambda_n(1+\lambda_n)^{-1}$ for $n \geq 1$. In (b), the capacity can be attained by a Gaussian signal process with covariance operator as above, with $u_n = Ue_n$ and
$\beta_n = (1+\lambda_n)^{-1}[\sum_1^K \lambda_i + P + K]/K$ for $n \leq K$;
$\beta_n = 0$ for $n > K$. In (c), the capacity cannot be attained.

A more general model is considered in [7]. That model is for any densely-defined linear operator, S. The capacity for the mismatched Gaussian channel can then be either smaller or larger than that of the matched channel, depending on the spectral properties of the operator S. The more general model is necessary in order to treat channels with jamming.

VII. CAPACITY OF SPHERICALLY-INVARIANT CHANNELS

The apparent usefulness of a spherically-invariant process to model noise in under-ice and shallow-water applications has motivated us to examine the channel capacity problem for communicating in such noise. This work has been aided by the work on signal detection described above; in fact, the likelihood ratio plays a key role in channel capacity problems.

We have examined the problem for the matched channel, where the constraint on transmitted signal is $E\|A(m)\|_N^2 \leq P$, with N the channel noise. As shown in [4] and [14], the capacity for the matched Gaussian channel with this constraint is $P/2$, with or without feedback. For the spherically-invariant channel with noise model $N_t = AG_t$, A a random variable independent of the Gaussian process (G_t), $EA^2 = 1$, we have found the capacity to be equal to $\frac{P}{2}E(A^{-2})$. $E(A^{-2})$ will typically be quite large for some underwater acoustics applications. Thus, this result holds forth the possibility that one may be able to communicate at much higher rates than for the Gaussian channel with the same covariance.

ACKNOWLEDGEMENT

Much of the work described in this summary [8,9,10] was done jointly with Professor A.F. Gualtierotti, IDHEAP, University of Lausanne, Switzerland.

REFERENCES

(1) C.R. Baker, "Optimum quadratic detection of a random vector in Gaussian noise," *IEEE Trans. on Communications Technology, 14,* 802-805 (1966).

(2) C.R. Baker, "On the deflection of a quadratic-linear test statistic," *IEEE Trans. on Information Theory, 15,* 16-21 (1969).

(3) C.R. Baker, "On equivalence of probability measures," *Annals of Probability, 1,* 690-698 (1973).

(4) C.R. Baker, "Capacity of the Gaussian channel without feedback," *Information and Control, 37,* 70-89 (1978).

(5) C.R. Baker, "Absolute continuity of measures on infinite-dimensional linear spaces," *Encyclopedia of Statistical Sciences, Vol. 1,* John Wiley & Sons, 3-11 (1982).

(6) C.R. Baker, "Channel models and their capacity," *Essays in Statistics: Contributions in Honour of Norman L. Johnson,* 1-16, P.K. Sen, ed. (North Holland, 1983).

(7) C.R. Baker, "Capacity of mismatched Gaussian channels," *IEEE Trans. on Information Theory* (to appear).

(8) C.R. Baker and A.F. Gualtierotti, "Signal detection and channel capacity for spherically-invariant processes," *Proceedings 23rd IEEE Conference on Decision and Control,* 1444-1446 (1984).

(9) C.R. Baker and A.F. Gualtierotti, "Discrimination with respect to a Gaussian process," *Probability Theory and Related Fields, 71,* 159-182 (1986).

(10) C.R. Baker and A.F. Gualtierotti, "Likelihood ratios and signal detection for non-Gaussian processes," *Stochastic Processes in Underwater Acoustics,* (Lecture Notes in Control and Information Sciences, v. 85), 154-180, C.R. Baker, ed. (Springer-Verlag, 1986).

(11) R.F. Dwyer, "A technique for improving detection and estimation of signals contaminated by under-ice noise," in *Statistical Signal Processing,* E.J. Wegman and J.G. Smith, eds., 153-166, Marcel Dekker (1984).

(12) S. Geman, "An application of the method of sieves: functional estimator of the drift of a diffusion," *Colloq. Math. Soc. Janos Bolyai,* 32, (Nonparametric Statistical Inference), Budapest (1980).

(13) T. Hida, "Canonical representations of Gaussian processes and their applications," *Mem. Coll. Science, Kyoto University,* 33A, 109-155 (1960).

(14) M. Hitsuda and S. Ihara, "Gaussian channels and the optimal coding," *J. Multivariate Analysis, 5,* 106-118 (1975).

(15) R.S. Liptser and A.N. Shiryayev, *Statistics of Random Processes I: General Theory,* Springer-Verlag (1977).

(16) G.R. Wilson and D.R. Powell, "Experimental and model density estimates of underwater acoustic returns," in *Statistical Signal Processing,* E.J. Wegman and J.G. Smith, eds., 223-240, Marcel Dekker (1984).

APPENDIX

Absolute Continuity and Likelihood Ratio

Definitions and Notation

All stochastic processes are defined on the probability space (Ω,β,P), with parameter set $[0,1]$. R^K is the Borel σ-field for R^K, $K<\infty$, $C_0[0,1] \equiv C_0$ is the set of all real-valued continuous functions on $[0,1]$ that vanish at zero. C is the Borel σ-field on C_0 defined by the sup norm. C^K is the Borel σ-field of C_0^K under the product topology; C_0^K can be identified with the set of all K-component real-valued vector functions having each component in C_0.

Suppose that (V_t) is a vector stochastic process such that $V(\omega,\cdot) \in C_0^K$ a.e. $dP(\omega)$. V will denote the corresponding path map from Ω into C_0^K, and P_V the induced measure on C^K:$P_V = P \circ V^{-1}$.

(N_t) will denote the noise; it is m.s.-continuous, Gaussian, zero-mean, and vanishes at t=0 w.p. 1. (N_t) is thus purely deterministic, so has a proper canonical Cramer-Hida representation [13]:

$$N_t = \sum_{i=1}^{M} \int_0^t F_i(t,s)dB_i(s) \tag{1}$$

where $M \leq \infty$ is the multiplicity of (N_t), each F_i is a deterministic Volterra kernel, and the B_i's are mutually orthogonal stochastic processes with orthogonal increments. (N_t) is Gaussian; the B_i's are thus mutually independent Gaussian processes with independent increments and continuous variances. Each B_i is thereby path-continuous. Since the representation (1) is proper canonical, and (N_t) and the family of B_i's are Gaussian, the σ-field generated by $\{N_u, u \leq s\}$ is the same as the σ-field generated by $\{B_i(u), u \leq s, i \leq M\}$, for all s in $[0,1]$. β_i will denote the Borel measure on $[0,1]$ defined by the continuous non-decreasing variances EB_i^2, $0 \leq s \leq 1$.

We assume that the multiplicity M of (N_t) is finite. This restriction is due to the absence of some needed results in infinite-dimensional stochastic calculus. However, based on a partial investigation, we believe that the results on absolute continuity and likelihood ratio presented here remain valid for $M=\infty$.

Suppose that (V_t) is any stochastic process; $\underline{a}(V)$ is the P-completed filtration generated by (V_t), and $\underline{a}(V) \vee \underline{a}(N)$ is the smallest filtration containing both $\underline{a}(V)$ and $\underline{a}(N)$. We recall that a process (X_t) is $\underline{a}(V)$-predictable if $G:(t,\omega) \rightarrow X_t(\omega)$ is measurable with respect to the predictable σ-field $P(V)$ in $\mathbb{R}^+ \times \Omega$; $P(V)$ is generated by all path-continuous stochastic processes that are adapted to $\underline{a}(V)$.

R_N will denote the covariance function of (N_t), H_N its RKHS (reproducing Kernel Hilbert space) with inner product $<\cdot,\cdot>_N$, and R_N the covariance operator of (N_t) in $L_2[0,1]$. Range $(R_N^{1/2})$ is a separable Hilbert space, isomorphic to H_N, under the inner product, $(u,g)_N = \sum_n <u,e_n><g,e_n>/\lambda_n$, where $<\cdot,\cdot>$ is the $L_2[0,1]$ inner product, $\{\lambda_n, n \geq 1\}$ are the non-zero eigenvalues of R_N, and $\{e_n, n \geq 1\}$ are associated o.n. eigenvectors.

$\mathbb{R}^{[0,1]}$ is the space of real-valued functions on $[0,1]$; $R^{[0,1]}$ is the Borel σ-field generated by the cylinder sets $\{f$ in $R^{[0,1]}$: $(f(t_1),\ldots,f(t_n)) \in A^n\}$, $n < \infty$, A^n a Borel set in $\mathbb{R}^n$.

For a scalar stochastic process (V_t), υ_V is the probability induced on $\mathbb{R}^{[0,1]}$ by (V_t). If (V_t) has paths belonging a.s. to $L_2[0,1]$, then μ_V will denote the probability induced by the path map on the Borel σ-field of $L_2[0,1]$. If υ_1 and υ_2 are two probabilities on the same σ-field, then $\upsilon_1 << \upsilon_2$ means that υ_1 is absolutely continuous with respect to υ_2.

Absolute Continuity

Theorem 1 [9]: Let (V_t) be a stochastic process independent of (N_t). Suppose that (Y_t) is a process such that $\upsilon_Y << \upsilon_N$.

If (Y_t) is adapted to $\underline{a}(N) \vee \underline{a}(V)$, then $Y_t = S_t + N_t^*$ a.e. dP for each fixed t in $[0,1]$, where (N_t^*) has the same finite dimensional distributions as (N_t), and is adapted to $\underline{a}(Y)$. $N_t^* = \sum_{i=1}^{M} \int_0^t F_i(t,s)dB_i^*(s)$ a.e. dP, each fixed t in $[0,1]$, where the B_i^*'s are mutually independent zero-mean Gaussian processes, (B_t^*) has the same law as (B_t), and $\underline{a}(B^*) = \underline{a}(N^*)$. Moreover,

$$S_t = \sum_{i=1}^{M} \int_0^t F_i(t,s)\phi_i(s)d\beta_i(s), \tag{2}$$

where $(\phi_i(t)), i \leq M$, is a stochastic process that is $\underline{a}(Y)$-predictable and has paths a.s. in $L_2[\beta_i]$. □

If both (N_t) and (Y_t) have continuous paths, then Theorem 1 can be strengthened. In that case, let P_N' and P_Y' be the induced measures on C. Then $\upsilon_Y << \upsilon_N <=> P_Y' << P_N' <=> \mu_Y << \mu_N$.

Theorem 2 [9]: Let (V_t) be a stochastic process independent of (N_t). Suppose that (S_t) is a stochastic process adapted to $\underline{a}(N) \vee \underline{a}(V)$ and with paths a.s. in H_N.

(1) If $X_t = S_t + N_t$ a.e. dP, for each fixed t in $[0,1]$, then $\upsilon_X << \upsilon_N$.

(2) If $X_t = S_t + N_t$ a.e. dtdP, then $\mu_X << \mu_N$. □

Likelihood Ratio

Suppose that (Y_t) satisfies the measurability assumption in Theorem 1, and that $\upsilon_Y << \upsilon_N$. Define a vector process (Z_t) with paths a.s. in C_0^M by

$$Z_i(t) = \int_0^t F_i(t,x)\phi_i(s)d\beta i(s) + B_i(t) \tag{3}$$

where ϕ_i is defined in Theorem 1. In this case, $P_Z << P_B$ [15].
Theorem 3 [9]: Suppose that (Y_t) satisfies the sufficient conditions of Theorem 2. Then

$$\frac{d\upsilon_Y}{d\upsilon_N}(x) = \int_{C_0^M} [dP_Z/dP_B](y)\; dP_B\rfloor_{N=x}\,(y)$$

a.e. $d\upsilon_N(x)$, where $P_B\rfloor_{N=x}$ is the conditional measure of B given $N = x$. If (S_t) is defined as in Theorem 1, and $Y = X + N$, then

$$\frac{d\mu_Y}{d\mu_N}(x) = \int_{C_0^M} [dP_Z/dP_B](y)\hat{P}(x,dy)$$

a.e. $d\mu_N(x)$, where $\hat{P}$ is a transition probability on $L_2[0.1] \times C^M$, and $\hat{P}(x,\cdot) \perp P_B$ a.e. $d\mu_N(x)$. Moreover, $\hat{P}(x,\cdot)$ is a point mass on C_0^M, giving probability one to $\{m(y)\}$, where

$$[m_i(y)](t) = \sum_n <y,e_n><f_t^i,e_n>/\lambda_n \text{with}$$

$$f_t^i(s) = \int_0^t F_i(s,u)d\beta_i(u).$$

□

From Theorem 3, one can obtain $d\upsilon_Y/d\upsilon_N$ and $d\mu_Y/d\mu_N$ from dP_Z/dP_B. Since (B_t) is a vector process with components that are mutually independent continuous-path Gaussian martingales w.r.t. $\underline{\alpha}(N)$, dP_Z/dP_B can be obtained from well-known results [15].

DETECTION IN A NON-GAUSSIAN ENVIRONMENT: WEAK AND FADING NARROWBAND SIGNALS

Stuart C. Schwartz and John B. Thomas
Department of Electrical Engineering
Princeton University
Princeton, New Jersey 08544

ABSTRACT

Procedures for the detection of both weak and narrowband signals in non-Gaussian noise environments are discussed. For the weak signal case, nonlinear processors based on the Middleton Class A noise model and the mixture representation are developed. Significant processing gains are achievable with some rather simple procedures. A variant of the mixture model leads to a nonstationary detector, called the "switched detector." Experiments with this detector on ambient arctic and shrimp noises show processing gains of 1.4 to 4.1 dB, respectively.

Signals with a moderate signal-to-noise ratio in non-Gaussian noise are modeled as fading narrowband signals. A new processor is developed which combines a robust estimator (for the fading signal) with a robust detection procedure. The robust estimator-detector preserves the structure of the quadrature (envelope) matched filter and is shown to be asymptotically optimal for a wide range of decision rules and several common target models encountered in sonar and radar. Various degrees of robustness are achieved, depending on the assumed availability of noise reference samples.

I. INTRODUCTION

It is widely acknowledged that signal detectors designed for Gaussian statistics may suffer significant degradation when the actual statistics deviate from the assumed model. Classical examples are the matched filter for coherent reception of deterministic signals and envelope processing for both incoherent reception and stochastic signals. For both of these examples, there is serious degradation with the noise actually a mixture density with a Gaussian nominal and minor contamination. A converse has also been noted: a modest degree of nonlinear processing can lead to a detector with considerable processing gain over the comparable linear detector or classical detector developed under the Gaussian noise assumption.

The purpose of this paper is to demonstrate some of the processing gains that are achievable by acknowledging and modeling non-Gaussian environments. We concentrate on two general areas - weak and narrowband signals. In the first, we focus on the detection of weak signals in non-Gaussian noise. We discuss the usefulness of the Middleton Class A narrow-band noise model in determining non-linear signal processors that closely approximate optimal procedures. Comparative performance is measured in terms of asymptotic relative efficiency (ARE) and substantial improvement over linear (Gaussian model) processing is indicated.

One of these approximations leads in a natural way to a non- stationary detector, termed a *switched detector*. Ambient arctic noise and shrimp noise are used as test cases for a number of these approximations and improvement in processing gain is measured to range from 1.4 to 4.1 dB.

The second major focus of this paper is the detection of fading narrowband signals in non-Gaussian noise. A robust procedure is developed for the classical envelope detector which clearly illustrates the penalty one must pay in detectability for maintaining robustness in false alarm. A new processor is then developed which combines a robust estimator (for the fading signal) with a robust detection procedure. While this robust estimator-detector has an asymptotic minimax property, interestingly, a set of Monte Carlo experiments shows that the asymptotic analysis accurately predicts the small sample performance - in some cases, down to 16 samples.

Since the narrowband fading signal model encompasses many realistic situations in radar and sonar, it is expected that the combined robust estimator-detector will find a number of applications in high resolution systems where non-Gaussian statistics can no longer be ignored.

II. THRESHOLD DETECTION IN NARROWBAND NON-GAUSSIAN NOISE

Among the physically motivated models for non-Gaussian noise probability density functions (pdf), Middleton has developed two general classes, depending on whether the noise could be considered narrowband or broadband. [1], [2]. For the narrowband case, or Class A noise, it is assumed that the spectrum of the noise is comparable or narrower than the receiver passband. In this study, we focus on this Class A model.

This model has been shown to be a very accurate first-order statistical description for a surprisingly wide variety of noise and interference situations, both man-made and natural. The general situation described is one in which there is an impulsive interference plus a background Gaussian component. The impulsive component consists of interference sources that are distributed in space and time according to Poisson statistics. A complicated statistical averaging over time epochs, Doppler velocities, etc... has to be performed. However, after some manipulation, the Class A pdf can be expressed as an infinite sum of weighted Gaussian densities [3]. The infinite series, with a normalized variance, contains only two free parameters which can be measured directly from the data [4]. Consequently, since the form of this sum is independent of the particular physical situation, the model is truly canonical - a distinct advantage of the Middleton approach.

The Class A pdf is given by

$$f(y) = \sum_{m=0}^{\infty} k_m \, g(y;0,\sigma_m^2) \tag{1}$$

where $g(y;0,\sigma_m^2)$ denotes the Gaussian pdf with mean 0 and variance σ_m^2, and

$$\sigma_m^2 = \frac{m/A + \Gamma}{1 + \Gamma} \tag{2}$$

$$k_m = e^{-A} \, A^m / m! \tag{3}$$

The two parameters mentioned above are A and Γ. The parameter Γ gives the ratio of the power in the Gaussian interference component to the power in the impulsive component. Small Γ means that the noise is basically impulsive - highly non-Gaussian. The parameter A is the product of the mean duration of a typical interfering signal and the mean arrival rate of the impulsive interferers. Large values of A imply interfering waveforms that overlap, with the statistics approaching Gaussian as A further increases. Small A implies impulsive noise with the associated pdf exhibiting larger tails. Graphs of pdf's for different values of A and Γ can be found in [1]-[3].

In a series of studies by Vastola, and Vastola and Schwartz, ([5]-[7]), it has been demonstrated that a truncated version of (1) can be a very good approximation to the infinite series. Indeed, with two or three, or at most four terms, one can not distinguish the approximation to the envelope distribution. Ref. [5], Figures 1-4, illustrate the distribution and approximations of the instantaneous envelope whose pdf is given by (1).[1]

This close approximation also carries over to the actual detector used in the weak signal case, i.e., the locally optimum detector. Thus, if one computes the expression

$$l_0(y) = -f'(y)/f(y) \tag{4}$$

using (1) and also using a truncated version, say two or three or four terms of the series, there is a very close agreement in the two curves. (See [5], Figs. 5-8.) For most practical purposes, the two curves coincide. One exception is in the tails of the locally optimum detector for moderate values of A ($A \approx 0.3$) and low values of Γ ($\Gamma \approx 0.0005$).

For the weak signal case, using an approximation to the locally optimum nonlinear detector leads to results which are essentially the same as the optimum (given by the ratio of two infinite sums) and significantly better than the linear detector. (The linear detector is, of course, both optimum and locally optimum under a Gaussian noise assumption.) For cases of large departures from Gaussian noise (i.e., $\Gamma = 0.0005$, $A = 0.3$), there can be up to a 30dB processing gain in terms of ARE over the linear detector.[2] For further details, see Table I, Ref. [5]. A parallel study was done for stochastic signals in non-Gaussian noise. Although the envelope distribution series expressions are obviously more complicated, the approximations with a few terms appear to converge very rapidly (Ref. [7], Figs. 5-7). Thus, we are led to believe that, for the weak stochastic signal case in non-Gaussian noise, significant processing gains can again be achieved by truncating the appropriate series.

It is noteworthy that the structure of the locally optimum detector nonlinearity suggests two other general approaches to the weak signal detection problem. The first is the use of other, more common, nonlinearities in the detector structure. The second is the use of switched detectors. Both will now be discussed.

A typical locally optimum detector looks like that given in Figure 1. By observing that most of the observations fall within regions A and C, it is suggested that the common "hole puncher", Fig. 2, will be a good approximation to the optimum nonlinearity. This was indeed verified in an evaluation of ARE for a number of values of Γ and A. In almost all cases, the ARE of the hole puncher was indeed close to the optimum ARE. Somewhat disappointing, however, was the degree of sensitivity of the hole puncher to errors in assumed parameter values. Indeed, in one study performed (see [5], Fig. 13), the optimum performance of the hole puncher deteriorated rapidly while the performance of a comparable soft-limiter was substantially more robust. A comparative analysis of three

[1] Ref. [7], Figures 2-4, give the envelope distribution for an additive stochastic signal in non-Gaussian noise. It is evident that similar conclusions can be reached in this more complicated signal detection problem.

[2] For the convenience of the reader, Appendix A summarizes the expression for asymptotic relative efficiency.

nonlinearities - hole puncher, soft-limiter, hard-limiter - was made to study further the issues of processing gain and sensitivity. (Refs. [5],[6].) The hole puncher had an ARE close to the optimum, and significantly higher than the ARE's of the other two common nonlinearities when the assumed model had the correct parameter values. As suggested above, however, its performance was sensitive to the accuracy of the model. Clearly, there is a need for further study of desensitizing the hole-puncher to errors in model parameters.

III. SWITCHED DETECTORS

As mentioned above, the shape of the nonlinearity of the locally optimum detector suggests both some common nonlinearities and also an approach we have called the *switched detector*.

From Figure 1, it is clear that since a small percentage of the observations fall in region B, most observations will be in regions A or C. Thus, these two linear regions can be represented by two linear detectors which are "switched", depending on the noise state, to process the observations.

There is yet another viewpoint that clearly illustrates the philosophy of switching between two linear detectors. If we assume that two terms of the pdf series in (1) are sufficient to represent accurately the noise pdf, then, after suitable normalization, we have

$$f(y) = (1-\epsilon)\, g(y;0,\sigma_0) + \epsilon\, g(y;0,\sigma_1) \tag{5}$$

where the mixing parameter is given by

$$\epsilon = \frac{A}{1+A} \tag{6}$$

and the ratio of variances is

$$\frac{\sigma_1^2}{\sigma_0^2} = 1 + \frac{1}{A\Gamma} \tag{7}$$

Equation (5) is the well-known Gaussian-Gaussian mixture. This is perhaps the most frequently used example in robustness studies of the more general mixture model where, typically, the second pdf in (5) is taken as an arbitrary pdf. (Oftentimes, this second pdf is taken to be heavy-tailed so as to represent impulsive noise.)

Now, if one takes the viewpoint that the noise is Gaussian, but with variance either given by σ_0 or σ_1, with $\sigma_1 >> \sigma_0$, then the mixing parameter ϵ represents the percentage of time the Gaussian noise has the larger variance σ_1. Consequently, the pdf, Eq. (5), is then exact as it represents the unconditional pdf of the noise.

Given that one is trying to detect a weak signal in Gaussian noise, the locally optimum detector is

$$l_0(y) = -g'(y;0,\sigma)/g(y;0,\sigma) = y/\sigma^2 \tag{8}$$

Thus, the detector is linear with slope given by the reciprocal of the variance of the noise. (See Figure 3.)

The two philosophies should now be clear. For mixture noise, one can use the locally optimum detector given by $-f'(y)/f(y)$, where $f(y)$ is given as in (5). The optimum weak signal detector then takes the form:

$$l_0(y) = -f'(y)/f(y) = y \left\{ \frac{\frac{(1-\epsilon)}{\sigma_0^2}\, g(y;0,\sigma_0) + \frac{\epsilon}{\sigma_1^2}\, g(y;0,\sigma_1)}{(1-\epsilon)\, g(y;0,\sigma_0) + \epsilon\, g(y;0,\sigma_1)} \right\} \tag{9}$$

This is a nonlinear function, denoted by g_ϵ in Figure 3.

The other point of view is to assume the noise is Gaussian, with standard deviation either σ_0 or σ_1. Then, one uses a linear detector, but has to decide first which noise pdf is governing the observation. Hence, here we have a "switched detector", switching between the two linear detectors given by g_0 and g_1 in Figure 3.

These two approaches were evaluated in a series of investigations ([8],[9]). Figure 4 gives a comparison of the performance of the switched detector and the locally optimum detector for Gaussian mixture noise, Eq. (9). Performance is evaluated in terms of ARE. It is clear that the switched detector outperforms the mixture nonlinearity. This is especially so in the mid-range of variances, $5 \le \sigma_1^2/\sigma_0^2 \le 100$ and moderate values of the mixing parameter ϵ. For $\epsilon = .3$, the gain is slightly below 1.5 dB (ARE=1.4).

The processing gain is substantially greater when comparing the switched detector to a fixed-structure linear detector; here the gain is of order $\epsilon\, \sigma_1^2/\sigma_0^2$. For variances and mixing parameter values quoted above, the processing gain is in the neighborhood of 12 dB.

These performance figures assume exact knowledge of the switching sequence $\{\epsilon_i\}$. In practice, of course, it will be necessary to estimate ϵ and errors will necessarily result, causing a degradation in the switched detector performance. Figures 5,6,7 show how this performance is affected as a function of the probability $p_{0/1}$ of incorrectly classifying an impulsive noise sample as a background noise sample and of the probability $p_{1/0}$ of incorrectly classifying a background noise sample as an impulsive noise sample. The most obvious conclusion to be drawn from these figures is the importance of recognizing, as quickly as possible, the inception of a burst or impulsive mode. Clearly, even a few unrecognized impulsive noise samples will seriously affect detector performance. Consequently, it is imperative that the first decision, which noise state the observations come from, be made with as much precision as possible.

A switching detector was implemented with a sample of Arctic under-ice noise ([8]). The data stream was processed in blocks, and a threshold test, set at 1.282 $\hat{\sigma}$ ($\hat{\sigma}$ = the estimated variance of the block), was used to determine which Gaussian noise was being observed.[3] For the 58 data blocks processed in this manner, the parameters were estimated to be $\hat{\sigma}_1^2/\hat{\sigma}_0^2 = 9.03$ and $\hat{\epsilon} = 0.089$. Thus the high variance noise is observed about 9% of the time. The ARE of the switched detector compared to a linear detector was also computed using Eq. (A.5). The estimate is 1.58, which is a processing gain of almost 2 dB. In the next experiment, performance was evaluated in terms of false alarm and power, rather than ARE.

For this second experiment, the philosophy of the switched-detector model was extended to non-Gaussian mixture models as well [10]. The underwater noise was produced by a concentration of shrimp in the Pacific Ocean near Hawaii; this noise sample (96000 data points) is characteristic of a highly non-Gaussian environment. The pdf has extremely heavy tails and approximately one out of ten thousand observations will have an amplitude greater than 16 σ. For this experiment, a nonparametric detector ([9]) was

[3] The value 1.282 $\hat{\sigma}$ corresponds to an error rate of $p_{1/0} = .2$ under the assumption of Gaussian noise.

used to identify the beginning of each burst period. Some preliminary data analysis suggested that the Johnson S_u family would be a better fit to shrimp noise. Consequently, this was the family used in the mixture pdf, and it produced a good fit to the empirical pdf's of the data set, both in "background" and "burst" modes. Johnson S_u noise can be obtained from a memoryless transformation $g(x)$ on unit-power Gaussian noise where

$$g(x) = \lambda \sinh(\frac{x}{\delta})$$

Since the parameter δ has more control over tail behavior than does λ, it will be called the *tail parameter*. Small values of δ give heavy tails and the distribution approaches Gaussian as $\delta \rightarrow \infty$.

Comparisons were made among the following detectors:

D1 - The linear detector given by (8)

D2 - The Gaussian-Gaussian mixture given by (9)

D3 - The Laplace noise detector, where the pdf is

$$f(y) = \frac{1}{\sqrt{2}\sigma} e^{-\frac{\sqrt{2}}{\sigma}|y|}$$

D4 - The Johnson noise detector, where the pdf is

$$f(y) = f(y;\delta,\lambda) = \frac{1}{\sqrt{2\pi}} \frac{\delta}{\lambda} \left[1 + (\frac{y-a}{\lambda})^2\right]^{-1/2} e^{-1/2\{\gamma + \delta \sinh^{-1}(\frac{y-a}{\lambda})\}^2}$$

The parameters δ and λ affect the shape and scale of the densities. The choice of

$$\lambda = \left[\frac{2\sigma^2}{e^{2/\delta^2} - 1}\right]^{1/2}$$

yields densities of common variance σ^2.

D5 - The Johnson switched detector, where switching is done between two pdfs of the Johnson family, using "high mode" parameters $\delta = 1.03$ and $\lambda = 67.1$. Observations judged to have come from the high mode were multiplied by one-third, the approximate ratio of the standard deviations of the two modes.

The results of the experiment are shown in Figure 8 where false-alarm and power are plotted on a logarithmic scale. These receiver operating curves suggest at least two preliminary conclusions. First, it is clear that the linear detector can perform quite poorly when the noise is decidedly non-Gaussian, such as in this situation. Secondly, it also appears that significant improvement in detector performance can be achieved with some nonlinear processing, e.g., a "Laplace noise" or "Johnson noise" detector. Furthermore, it would appear that substantial improvements can be obtained in a non-Gaussian, non- stationary environment by judicious mode switching. Thus, it would appear fruitful to study signal detection procedures in nonstationary environments as a two-stage decision procedure. First, one decides on the noise state that is governing the observations. After that decision is made, the optimum likelihood (or more general statistical) processor is used to do the signal detection and extraction. It is also clear that, as suggested above, system design should emphasize that the first decision (noise state) be made with high precision.

IV. DETECTION OF FADING NARROWBAND SIGNALS IN NON-GAUSSIAN NOISE

Although optimum statistical procedures are especially desirable in the low SNR or weak signal situation, substantial increases in system performance can also be achieved when there is a moderate SNR. This is especially evident when the assumed model deviates from the actual statistical model - a situation often encountered in non-Gaussian and non-stationary environments.

In this section, we consider a class of models that is applicable for a variety of radar and sonar systems - narrowband fading signals. Of particular interest is the detection of these signals in non-Gaussian noise and, more generally, over a range of SNR conditions. We first discuss robustifying the classical quadrature matched filter test in a natural way - with a limiter on the envelope statistic. It is shown that robustness in false alarm comes with a large degradation in the power of the test as the contamination (uncertainty) in the mixture noise distribution increases.

This unacceptable degradation is largely overcome with the introduction of a new approach. The procedure preserves the structure of the quadrature matched filter, which is the uniformly most powerful (UMP) test for Gaussian noise. In the new procedure, the usual sample-mean estimators are replaced by robust α-trimmed (minimax) estimators. The resulting test is asymptotically optimal for a wide family of decision rules and for several common target models often encountered in radar and sonar systems. The test is extended to handle the important case of unknown power level. Various degrees of robustness are achieved, depending on the assumed availability of noise reference samples.

We consider the model

$$z_i(t) = A\ s_i\ \cos(\omega_0 t + \theta) + n(t), \quad i = 1,2,...,M \tag{10}$$

where s_i is a known, positive amplitude modulation, A and θ are random amplitude and uniformly distributed phase, and $n(t)$ is narrowband non-Gaussian noise. The slow fading is represented by A being a random amplitude, independent of time. (Fast fading would require a random process, i.e., $A(t)$ rather than the constant A.) We will want to study a binary test of signal present or noise alone:

$$H_0\colon\ A = 0$$

$$H_1\colon\ A > 0$$

For the Gaussian noise case, the uniformly most powerful test (UMP) is the coherent sample envelope, $R(\mathbf{x},\mathbf{y})$, compared to a threshold. Thus, one takes the received signal $z_i(t)$, passes it through in-phase and quadrature channels and forms the coherent sample envelope:

$$R(\mathbf{x},\mathbf{y}) = (\sum_{i=1}^{M} x_i s_i)^2 + (\sum_{i=1}^{M} y_i s_i)^2 \tag{11}$$

where the in-phase and quadrature lowpass outputs are

$$x_i = \int_0^T z_i(t) \cos \omega_0 t \; dt = A \; s_i \cos \phi + n_{c_i} \tag{12}$$

$$y_i = \int_0^T z_i(t) \sin \omega_0 t \; dt = A \; s_i \sin \phi + n_{s_i} \tag{13}$$

Observe that this is UMP, independent of A being random or deterministic. This is a well-studied test, especially for incoherent reception, i.e., when θ is uniformly distributed. See [11], Chapter 5.

Unfortunately, even though this test has a UMP property, performance is quite sensitive to the noise assumption. For example, if the additive noise is not Gaussian, but of the mixture type with nominal Gaussian pdf, there is a substantial decrease in detection probability and a significant deterioration in false alarm rate. (See [12], Figures 2.1-2.5, [13]). One example which illustrates this sensitivity of the detector to the noise assumption is where the designer sets the detection threshold so as to achieve a 10^{-6} false alarm rate. This desired false alarm, in actuality, increases to 3×10^{-2} when the assumed Gaussian noise becomes a Gaussian-Gaussian mixture with only 10% contamination!

Clearly there is a need to de-sensitize or robustify the envelope detector. Since limiters have been used effectively to develop robust procedures in coherent reception, one naturally thinks of using a limiter on the envelope test. Thus, one takes as a model, Gaussian contaminated quadrature noise, and investigates the test obtained from considering a least favorable distribution on the envelope. This was done in [12], Chapter 3. The result was a limiter-type test of the form

$$T(R) = \begin{cases} c_1 , & R > c_1 \\ R , & c_2 < R < c_1 \\ c_2 , & R < c_2 \end{cases} \tag{14}$$

with R given by Eq.(11). The limiter break points are found in the usual fashion.

An analysis of this test shows that it is robust in terms of false alarm for increases in contamination of the mixture model, but only at the expense of a significant degradation in detectability. A typical result is shown in Figure 9, which illustrates the significant drop in power as the contamination is increased. Further details are given in [12], [16].

It is clear one wants to maintain a respectable detectability (power curve) while at the same time achieving a robustness of false alarm. This was the subject of a series of studies, [12], [14]-[16] where a new procedure was developed. This new approach combines a robust estimator for the fading signal with a robust quadrature detector. One version has an asymptotic minimax property and computer simulations are encouraging for all versions of the combined robust estimation-detection procedure.

Consider first the following test when noise-reference samples are available. Announce H_1, if

$$R_\alpha(\mathbf{x},\mathbf{y}) \geq t_{M'} \; W_\alpha(\mathbf{u},\mathbf{v}) \tag{15}$$

where

$$R_\alpha(\mathbf{x},\mathbf{y}) = \overline{X}_\alpha^2 + \overline{Y}_\alpha^2 \tag{16}$$

$$W_\alpha(\mathbf{u},\mathbf{v}) = \frac{1}{M'} \sum_{j=1}^{M'} W_{j\alpha}(\mathbf{u},\mathbf{v}) \tag{17}$$

$$W_{j\alpha}(u,v) = \overline{U}_{j\alpha}^2 + \overline{V}_{j\alpha}^2 \tag{18}$$

The superscript bar, i.e., $\overline{X}_\alpha$ denotes a minimax robust estimate of $A \cos \phi$. Here, we take Tukey's α-trimmed estimator

$$\overline{X}_\alpha = \frac{1}{N(1-2\alpha)} \left[\sum_{i=1+[N\alpha]}^{n-[N\alpha]} x_{(i)} \right] \tag{19}$$

where the $x_{(i)}$ are the ordered samples, $x_{(1)} \leq x_{(2)} \leq \cdots \leq x_{(N-1)} \leq x_{(N)}$ and $[N\alpha]$ is the greatest integer in $[0,N\alpha]$. α is the trim factor, usually taken in the range 0, 0.5.

The u,v symbols represent the noise reference samples which go through the same processing as x and y to obtain W. In search systems, these noise samples are easily obtained from adjacent resolution cells in range, doppler or bearing. Such systems are known as cell-averaging constant false alarm rate detectors (CA-CFAR). They are also known as mean level detectors (MLD) or sliding window detectors (SWD).

The minimax property relates to the marginal pdf of the noises n_{c_i}, n_{s_i}, which are assumed to be from the Tukey-Huber mixture

$$f(z) = (1-\epsilon)\, g(z;0,\sigma) + \epsilon\, c(z) \tag{20}$$

Observe that the nominal scale parameter σ will be taken as unknown, along with the pdf $c(z)$. At the end of this section, we will also discuss an extension of the procedure when the mixing parameter is assumed unknown. The random amplitude A is Rayleigh distributed

$$f(A) = 2A \; exp(-A^2/\overline{A}^2) \tag{21}$$

and, of course, the phase is uniformly distributed on $(0,2\pi)$.

Define the pdf

$$f_0^*(z) = (1-\epsilon)\, g(z;0,\sigma), \quad |z| < k\sigma$$

$$= \frac{(1-\epsilon)}{\sqrt{2\pi}\sigma} \exp\left(\frac{k^2}{2} - \frac{k\,|z|}{\sigma}\right), \quad |z| \geq k\,\sigma \tag{22}$$

where $k(\epsilon)$ is defined implicitly

$$\frac{g(k(\epsilon);0,1)}{k(\epsilon)} - \Phi(-k(\epsilon)) = \frac{\epsilon}{2(1-\epsilon)} \tag{23}$$

and Φ is the standard normal cumulative distribution function.

For a wide class of decision rules based on the family of translation invariant estimators, the test described above in Eqs. (15)-(19), and denoted by $d(\alpha)$, has a number of desirable asymptotic properties. They are listed below. Details, precise statements, and proofs can be found in [12], [16], which are available upon request. Basically, the proof relies on the equivalence of the asymptotic distributions of the various scale- invariant estimators, and on the existence of a saddlepoint for the estimation variance with the pdf $f_0^*(z)$, Eq.(22). Furthermore, we note that $W(\mathbf{u},\mathbf{v})$, is a consistent estimate of the variance σ.

As $M \rightarrow \infty$, the important asymptotic properties are:

1. $d(\alpha)$ is CFAR distribution-free and the false alarm is given by

$$P_{fa} = (1 + t_{M'}/M')^{-M'} \tag{24}$$

2. Set

$$\alpha^*(\epsilon) = \int_{-\infty}^{-k(\epsilon)} f_0(z)\, dz. \tag{25}$$

Then, f_0^* and $d^*(\alpha)$ are a saddlepoint pair:

$$\beta_{d \in D}(f_0^*,d) \leq \beta(f_0^*,d^*) \leq \beta_{f \in F}(f,d^*) \tag{26}$$

This is the desired maximin robust property. It puts a floor on the power for all pdf's $f(z)$ in the class, while maximizing the lower bound for all allowable test procedures. The false alarm is given by Eq.(24) and the power is

$$\beta(f_0^*,d^*) = \left[1 + \frac{t_{M'}}{M'(1 + S(f_0^*,d^*))}\right]^{-M'} \tag{27}$$

where the integrated SNR is

$$S(f_0^*,d^*) = \frac{M\overline{A^2}}{2V(f_0^*,\alpha^*)} \tag{28}$$

and

$$V(f,\alpha) = \frac{2}{(1-2\alpha)^2}\left\{\int_0^{F^{-1}(1-\alpha)} x^2 f(x)\, dx + \alpha(F^{-1}(1-\alpha))^2\right\} \tag{29}$$

3. As the number of reference samples increases, $M' \to \infty$, the test d^* coincides with the UMP test for the pdf f_0^*, Eq.(22), with σ known.

Although these results (items 1 and 2) are valid for $M \to \infty$, the asymptotic analysis accurately predicts detector performance for finite number of samples. Somewhat surprisingly, the agreement is excellent -- in some cases down to 16 samples.

Figure 10 gives representative results. Detection probability is plotted against effective integrated SNR which is

$$SNR_e = \frac{M\ E(A^2)}{\sigma_0^2}$$

where σ_0^2 is the variance of the nominal Gaussian pdf. False alarm appears as a parameter. Six cases are indicated for each SNR and false alarm. These include two analytical predictions from the asymptotic analysis and Monte Carlo simulations with four noise mixture pdf's. The two predictions are for $\epsilon = 0$, which is the usual sample mean unrobustified detector, and for f^* which is the maximin lower bound, i.e., the detector with the least favorable pdf. These two performance curves bracket the performance of the four mixture pdf's which are:

f_1: $\epsilon = 0$, the nominal Gaussian noise case

f_2: $\epsilon = 0.1$ and point masses at ± 15

f_3: $\epsilon = 0.1$ and Gaussian contamination with variance = 100

f_4: $\epsilon = 0.01$ and Gaussian contamination with variance = 900.

In all cases, the performance of the detector for finite samples is well approximated by the asymptotic analysis. The second observation is that detectability is relatively insensitive to changes in the noise pdf. For the contaminations studied, the maximum change in power is about 0.15, which is relatively modest, considering the wide range of contaminating noise pdf's. Put another way, for a fixed false alarm of 10^{-6}, the SNR loss to maintain a power of 0.5 or 0.9 is about 2 dB. The corresponding SNR loss for the classical UMP detector is 5 dB for $P_d = 0.5$ and 14 dB for $P_d = 0.9$. The lack of sensitivity of this procedure is also seen in the false alarm calculations for the same set of simulation parameters. This is given in Figure 11, where the curve marked T is that predicted by the asymptotic analysis, i.e., Eq. (24). The results indicate an "almost CFAR" detector for 16 samples!

These results illustrate that we have succeeded in robustifying the classical envelope detector. The penalty is in the complexity of computation. The classical procedure, which is usually implemented by a bank of FFT processors, is now replaced with a detector which has to sort and rank samples. This could be a formidable task at high bandwidths.

When reference samples are not available, or neighboring cells do not have the same statistics, the test, Eq. (15), has to be modified. The estimate of the scale (variance of the nominal pdf) and threshold value are now obtained from the same observations used to derive the test statistic. One test procedure is to use

$$R_\alpha(\mathbf{x},\mathbf{y}) \lessgtr \frac{t}{M} \left[S^2(\mathbf{x}) + S^2(\mathbf{y}) \right] \tag{30}$$

where R_α is given as before by Eq. (16) and

$$S(\mathbf{x}) = \text{median} \left\{ | x_i - \text{median} \{ x_i \} | \right\} \tag{31}$$

which is a variant of the M-estimate. Since we do not now have a consistent scale estimate (over the permissible mixture class), it has not been possible to prove a maximin result. On the other hand, the variance of the scale estimate is reasonably robust, and simulations using this test give performance which is close to results obtained previously.

Figure 12 summarizes a Monte Carlo simulation for this test, using the same parameters as before. The asymptotic analysis no longer brackets the finite sample case with $M = 16$. We do note, however, that the asymptotics will give close agreement for $M \geq 50$. (See [12].) Even for the case of 16 samples, however, the detector is robust in the sense that curves for the four mixture cases are close to one another. The variability in SNR, with $P_{fa} = 10^{-2}$ and 10^{-6}, is about 1.8 and 5.8 dB, respectively. (See [16], Section 5.4 for additional details.)

Other extensions to this combined estimation-detection procedure are possible. For the case of unknown Doppler frequency, one can use a bank of the above tests, similar to the procedure one would use in the classical envelope detector with unknown frequency. It turns out that, asymptotically, the loss incurred for straddling frequencies is the same as one encounters in the classical FFT implementation.

Finally, it is worth mentioning how one can handle uncertainties in the mixing parameter ϵ. In realistic situations, ϵ is unknown, but it may be possible to bound it:

$$\epsilon \leq \epsilon_{\max}$$

Then the mixture pdf Eq. (20) can be written as:

$$\begin{aligned} f(z) &= (1-\epsilon)\, g(z;0,\sigma) + \epsilon\, c(z) \\ &= (1-\epsilon_{\max})\, g + \epsilon_{\max} \left[(1 - \frac{\epsilon}{\epsilon_{\max}})\, g + \frac{\epsilon}{\epsilon_{\max}}\, c \right] \\ &= (1-\epsilon_{\max})\, g + \epsilon_{\max}\, c'(z) \end{aligned} \tag{32}$$

which is simply another mixture pdf. Consequently, one can then proceed to implement the tests discussed above using, instead, $\epsilon_{\max}$. The test will be robust for all $\epsilon \leq \epsilon_{\max}$, but the resulting bound on the variance will not be as tight as if ϵ were known. In computations for a range of values for ϵ, and various Gaussian-Gaussian mixtures, the efficiency loss was no greater than about 28% ([12], Section 5.1). Consequently, an increase of about 1.1 dB in SNR is needed to achieve the same detectability when confronted with uncertainty in ϵ. This is certainly a modest penalty to pay!

REFERENCES

1. Middleton, D., "Statistical-Physical Models of Electromagnetic Interference," *IEEE Trans. Electromgn. Compat.*, Vol. EMC-19, pp. 106-127, August 1977.

2. --- "Canonical non-Gaussian Noise Models: Their Implications for Measurement and for Prediction of Receiver Performance," *IEEE Trans. Electromgn. Compat.*, Vol. EMC-21, pp. 209-220, August 1979.

3. Spaulding, A.D. and Middleton, D., "Optimum Reception in an Impulsive Interference Environment - Part I: Coherent Detection," *IEEE Trans. Comm.*, Vol. COM-25, No. 9, pp. 910-923, September 1977.

4. Berry, L.A., "Understanding Middleton's Canonical Formula for Class A Noise," *IEEE Trans. Electromgn. Compat.*, Vol. EMC-23, No. 4, pp. 337-343, November 1981.

5. Vastola, K.S., "Threshold Detection in Narrow-Band Non-Gaussian Noise," *IEEE Trans. Comm.*, Vol. COM-32, No. 2, pp. 134-139, February 1984.

6. Vastola, K.S. and Schwartz, S.C., "Suboptimal Threshold Detection in Narrowband Non-Gaussian Noise," *Proc., IEEE Intl. Conf. on Comm.*, Boston, MA, pp. 1608-1612, June 1983.

7. Schwartz, S.C. and Vastola, K.S., "Detection of Stochastic Signals in Narrowband Non-Gaussian Noise," *Proc. IEEE Conf. on Decision and Control,* San Antonio, TX,

pp. 1106-1109, December 1983.

8. Czarnecki, S.V. and Thomas, J.B., *Nearly Optimal Detection of Signals in Non-Gaussian Noise,* T.R. #14, Dept. of Electrical Engineering, Princeton University, Princeton, NJ, February 1984.

9. Czarnecki, S.V. and Thomas, J.B., "Signal Detection in Bursts of Impulsive Noise," *Proc., 1983 Conference on Information Sciences and Systems,* Johns Hopkins University, Baltimore, MD, pp. 212-217, March 1983.

10. Willett, P.K. and Thomas, J.B., "The Analysis of Some Undersea Noise with Applications to Detection," *Proc., 24th Annual Allerton Conference on Communication, Control, and Computing,* Urbana, IL, pp. 266-275, October 1986.

11. Helstrom, C.W., *Statistical Theory of Signal Detection,* Oxford, Pergamon Press, 1968.

12. Weiss, M. and Schwartz, S.C., *Robust Detection of Fading Narrow-Band Signals in Non-Gaussian Noise,* T.R. #17, Dept. of Electrical Engineering, Princeton University, Princeton, NJ, February 1985.

13. Shin, J.G. and Kassam, S.A., "Robust Detection for Narrowband Signals in non-Gaussian Noise," *J. Acoust. Soc. America,* Vol. 74, pp. 527-533, August 1983.

14. Weiss, M. and Schwartz, S.C., "Robust Detection of Coherent Radar Signals in Nearly Gaussian Noise," *Proc., IEEE Intl. Radar Conference,* Arlington, VA, pp. 297-302, May 6-9, 1985.

15. Weiss, M. and Schwartz, S.C., "Robust Scale Invariant Detection of Coherent Narrowband Signals in Nearly Gaussian Noise," *Proc., IEEE Intl. Conf. on Acoustics, Speech, and Signal Processing,* Tampa, FL, pp. 1281-1284, March 1985.

16. Weiss, M. and Schwartz, S.C., *Robust and Nonparametric Detection of Fading Narrowband Signals,* T.R. #19, Dept. of Electrical Engineering, Princeton University, Princeton, NJ, January 1986.

APPENDIX A

Asymptotic Relative Efficiency

The locally optimum detector maximizes the slope of the power curve with respect to signal strength (evaluated at zero signal strength) for a fixed false alarm. Let $M_g(\alpha,\beta,\theta)$ be the number of observations necessary for the detector g to achieve the false alarm α and power β with signal strength θ. Then, a convenient measure for comparing two detectors in this weak signal situation is the asymptotic relative efficiency (ARE). It is defined as

$$ARE_{g_1,g_2} = \lim_{M_1 \to \infty} \frac{M_{g_2}(\alpha,\beta,\theta)}{M_{g_1}(\alpha,\beta,\theta)} \qquad \text{(A.1)}$$

Thus, ARE is the asymptotic ratio of the number of samples required for both detectors to achieve the same performance, as measured in terms of α and β.

To evaluate ARE, it is more convenient to use another expression. To this end, define the efficacy as

$$\eta_f(g) = \frac{\left[\int_{-\infty}^{+\infty} g'(x)\, f(x)\, dx \right]^2}{\int_{-\infty}^{+\infty} g^2(x)\, f(x)\, dx} \qquad \text{(A.2)}$$

where $f(x)$ is the noise pdf and $g(x)$ represents the detector nonlinearity used to process the observations in the form

$$l(x) = \sum_{i=1}^{M} g(x_i) \qquad \text{(A.3)}$$

That is, we assume the observations are i.i.d. and the processor passes each observation through the nonlinearity g, sums the output, and compares the result to a decision threshold.

The nonlinearity that maximizes the efficacy is

$$g(x) = -f'(x)/f(x) \qquad \text{(A.4)}$$

which happens also to be the Neyman-Pearson locally optimum procedure.

Under mild regularity conditions, it can be shown that the ARE is also given in the limit by ([11], Section VII.2):

$$ARE_{g_1,g_2} = \frac{\eta_f(g_1)}{\eta_f(g_2)} \qquad \text{(A.5)}$$

A close examination of (A.2) shows that efficacy, as a ratio of two expectations, can also be interpreted as an incremental SNR between two hypotheses. Consequently, ARE, as a ratio of efficacies or incremental SNR's, can be viewed as a processing gain in using one detector over another.

Equations (A.5) and (A.2) are more convenient expressions to compute ARE, especially when real data is being used as discussed in the text.

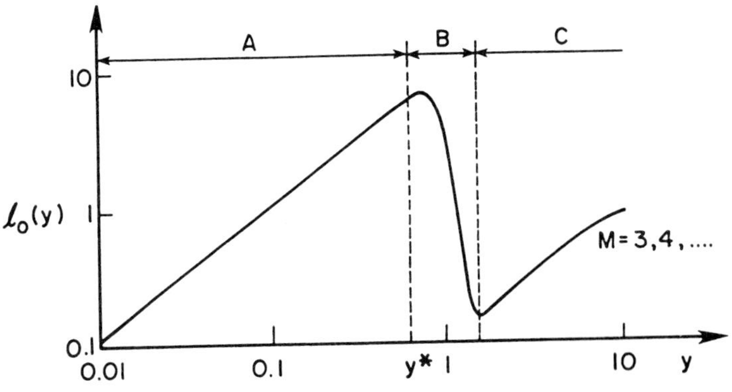

Figure 1 - Locally optimum detector nonlinearity for noise pdf $f_M, M = 2,3,4, \cdots$ (A=0.1,Γ'=0.1).

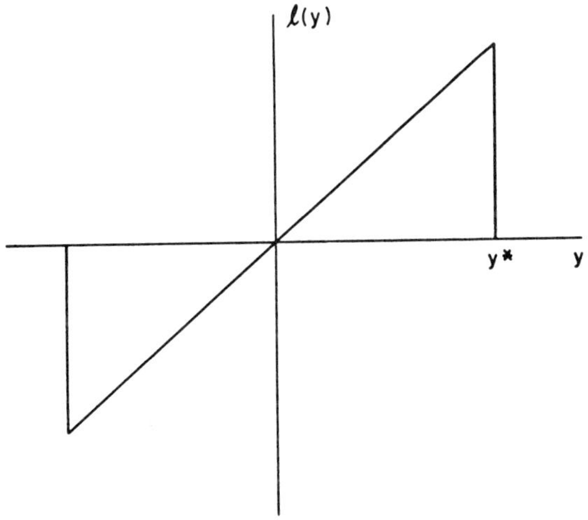

Figure 2 - Hole puncher as a nonlinear detector

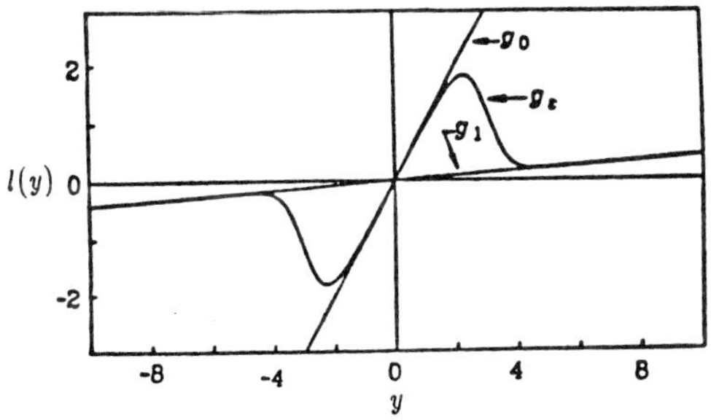

Figure 3 - The nonlinearity g_ϵ compared to the two linear detectors g_0 and g_1. The slopes of g_0 and g_1 are σ_0^{-2} and σ_1^{-2}, respectively.

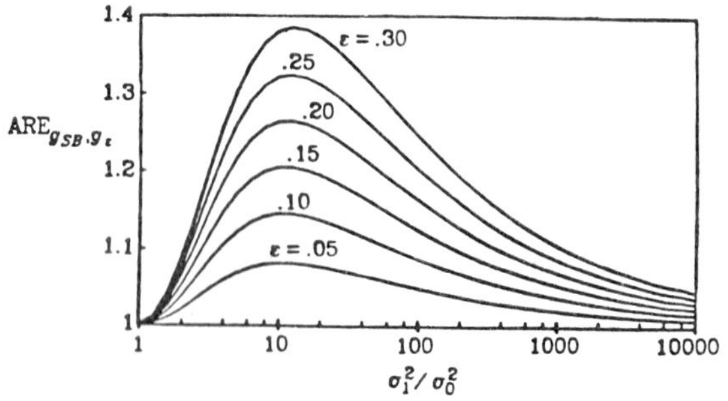

Figure 4 - Performance comparison of fixed nonlinearity g_ϵ and switched nonlinearity g_{SB} in Gaussian-Gaussian switched burst noise for various values of ϵ and range of σ_1^2/σ_0^2.

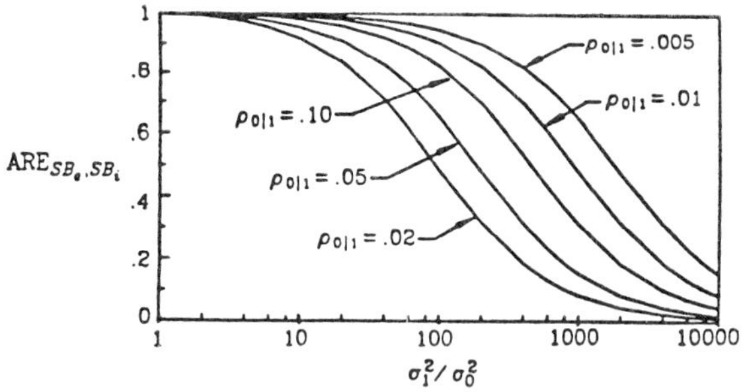

Figure 5 - Performance of switched detector with errors SB_e relative to ideal switched detector SB_i for various probabilities $\rho_{0|1}$ of incorrectly classifying an impulsive noise sample as background noise.

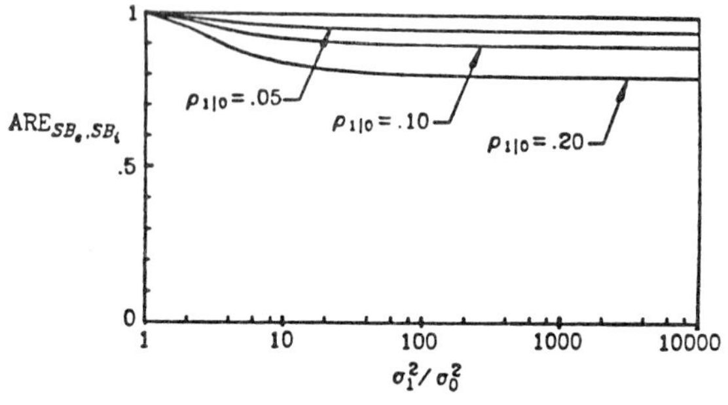

Figure 6 - Performance of switched detector with errors SB_e relative to ideal switched detector SB_i for various probabilities $\rho_{1|0}$ of incorrectly classifying a background noise sample as impulsive.

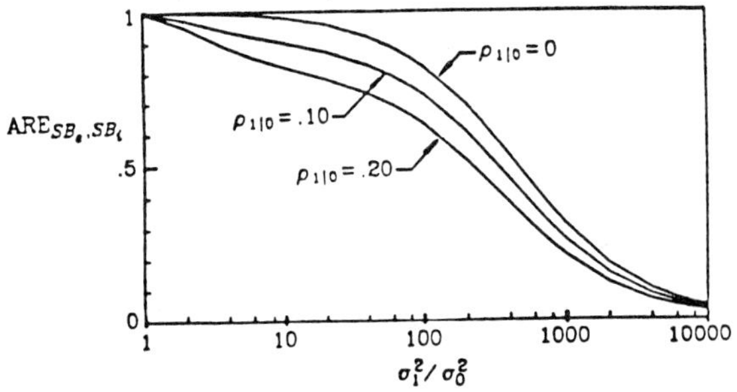

Figure 7 - Performance of switched detector with errors SB_e relative to ideal switched detector SB_i for various probabilities $\rho_{0|1}$ of incorrectly classifying a background noise sample as impulsive with fixed probability $\rho_{0|1}$=.02 of classifying an impulsive noise sample as background noise.

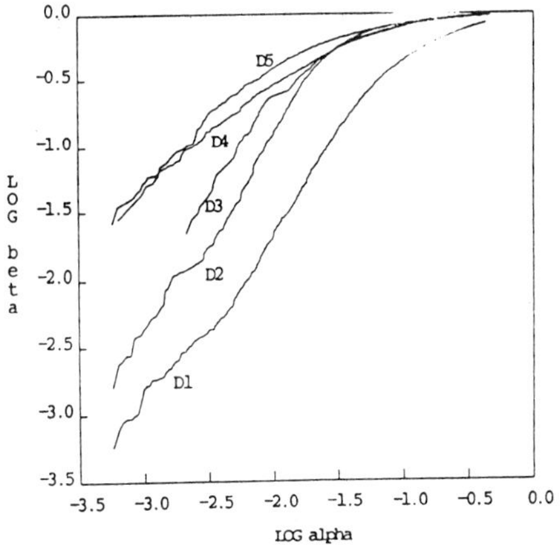

Figure 8 - A comparison of optimal and suboptimal detector performance in shrimp noise.

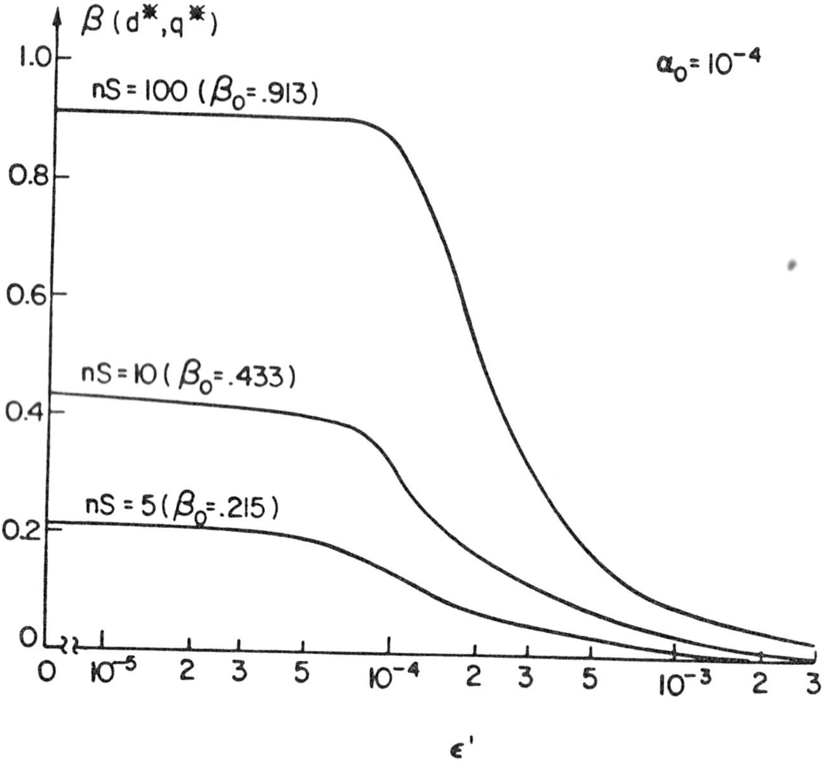

Figure 9 - Maximin bound on the detection probability vs. ϵ. $\alpha_0 = 10^{-4}$

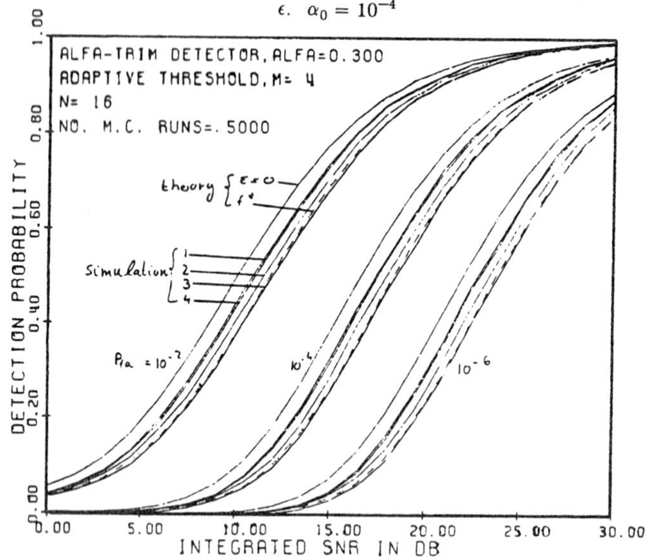

Figure 10 - Detection probability. Test - D1, $\alpha = .3$, $\Delta f = 0$

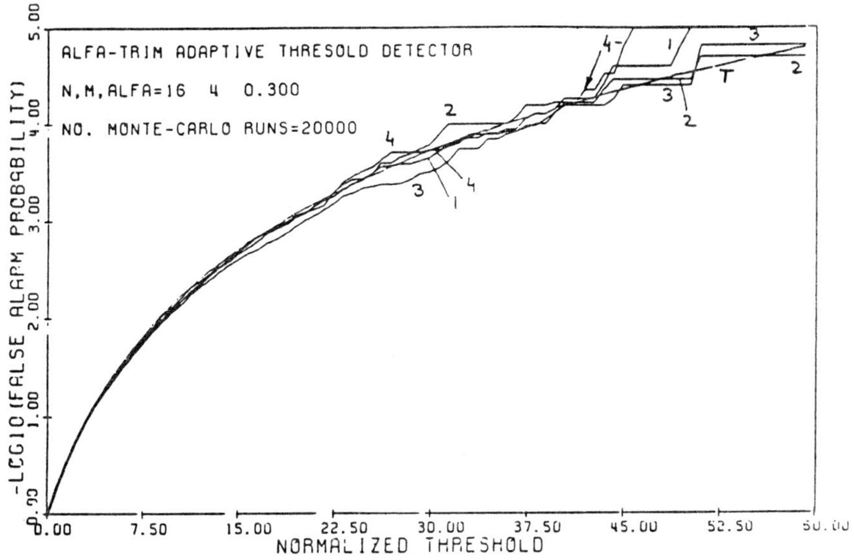

Figure 11 - False alarm probability. Test - D1, $\alpha = .3$.

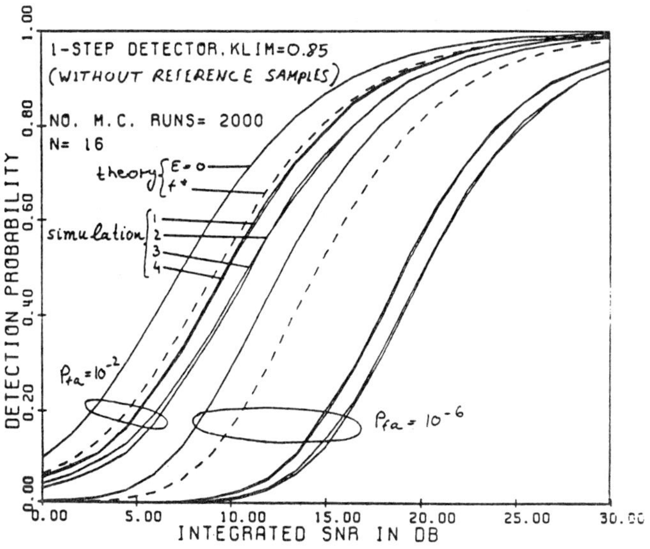

Figure 12 - Detection probability. Test - D4, $n = 16$, $\Delta f = 0$

ENERGY DETECTION IN THE OCEAN ACOUSTIC ENVIRONMENT

Fredrick W. Machell and Clark S. Penrod
Applied Research Laboratories
The University of Texas at Austin
Austin, Texas 78713-8029

ABSTRACT

The performance of the energy detector is evaluated using ambient noise data from several ocean acoustic environments. Estimates of the false alarm probability are presented as a function of the detection threshold for each environment. Estimated values for the corresponding minimum detectable signal-to-noise ratio (MDS) are also given for an artifically generated white Gaussian signal. The results presented here indicate that non-Gaussian noise statistics can have a significant impact on the relationship between the false alarm probability and the detection threshold. This threshold adjustment results in a serious degradation of energy detector performance in terms of the MDS for some non-Gaussian noise environments.

I. INTRODUCTION

In underwater acoustic signal detection, the noise process is usually assumed to be a Gaussian process (Ref. 1). This assumption is justified in many situations since a Gaussian noise field naturally arises in the ocean environment (Ref. 2). The Central Limit Theorem is often cited in attempt to strengthen the argument for a Gaussian assumption. However, statistical measurements of ocean acoustic ambient noise data indicate that noise statistics may deviate significantly from Gaussian in some environments (Ref. 3-6).

The purpose of this work is to examine the impact of non-Gaussian noise in the ocean environment on the performance of detectors designed under the Gaussian assumption. In particular, the performance of the energy detector is examined for several actual non-Gaussian ocean environments. These results are intended to give an indication of the performance degradation that might be expected with an energy detector in non-Gaussian environments. First, a more complete description of the detection problem under consideration is given. A discussion of the procedures used for evaluating detector performance is provided. Finally, these procedures are applied to the evaluation of the energy detector in the ocean acoustic environment.

II. THE DETECTION PROBLEM

Consider the following discrete time detection problem. Given the data or observation vector $X=(X_1,X_2, \ldots, X_m)$ decide between:

$$H_0\text{: } X \text{ is Gaussian, } N(0, \textstyle\sum_n)$$

$$H_1\text{: } X \text{ is Gaussian, } N(0, \textstyle\sum_{s+n})$$

The null hypothesis, H_0, states that the observation vector is a realization of a zero mean Gaussian process with covariance matrix $\sum_n$ (noise alone case). The alternative hypothesis, H_1, states that the observation vector is a realization of a zero mean Gaussian process with covariance matrix $\sum_{s+n}$ (signal + noise case). Note that the zero mean assumption is without loss of generality since the problem with non-zero mean can be transformed into an equivalent zero mean problem.

For this problem, the Neyman-Pearson nonlinearity is given by

$$g_{NP}(X) = X^T(\textstyle\sum_n^{-1} - \sum_{s+n}^{-1})X$$

This nonlinearity is known to maximize the probability of detection, $\beta = P(H_1 \mid H_1)$, for any detector with false alarm probability, $\alpha = P(H_1 \mid H_0)$. A special case of interest results when the samples are independent and identically distributed (i.i.d.) random variables. For this case, the nonlinearity reduces to the energy detector

$$g_{ED}(X) = \sum_{i=1}^{m} X_i^2$$

Several motivating factors can be provided for evaluating the performance of the energy detector in the ocean environment. The first concern is the establishment of a baseline for comparison with other detectors. The energy detector is also one of few detectors for which the theoretical performance is completely known (i.e., closed-form expressions for the distribution of the test statistic under both hypotheses). Furthermore, the detector nonlinearity does not depend on the signal-to-noise ratio (SNR). Energy detectors are also used in conjunction with FFT processing and beamforming algorithms. These schemes essentially use bandpass filtering in the frequency or spatial domains in attempt to improve the input SNR to the energy detector.

III. EVALUATION PROCEDURE

In the evaluation procedure, several factors need to be taken into consideration. First of all, since the energy detector is designed for i.i.d. samples, a sampling rate consistent with this assumption should be selected. From a theoretical point of view, with non-Gaussian data it may not be possible to guarantee that the samples are completely mutually independent. In such case, uncorrelated samples may be the best one can hope to obtain. Correlation function estimates can be used to check the degree of correlation between the samples.

Perhaps of more importance is choosing the criteria for measuring detector performance. One criterion considered here is the minimum detectable signal-to-noise ratio (MDS). The MDS is the minimum SNR required to achieve 50% probability of detection at a specified false alarm probability, and as such provides a measure of detectability of a signal. A related parameter of interest is the relationship between the false alarm probability and the detection threshold. Indeed, the adjustment of the threshold required to maintain a desired false alarm probability has a direct impact on the probability of detection in non- Gaussian environments. These parameters will be estimated for the energy detector in several environments.

For estimation of the false alarm and detection probabilities, the detector outputs are assumed to represent independent Bernoulli trials. This assumption is particularly reasonable if the samples entering the detector are uncorrelated. The general estimation procedure involves use of the Bernoulli distribution to obtain bounds on the false alarm and detection probabilities at a prescribed confidence level. The estimate at the 50% confidence level is an unbiased estimate and will be used here. Thus, the false alarm estimate is simply the proportion of noise alone detector outputs exceeding the threshold (see Refs. 7-8).

An estimate of the MDS may be obtained in a similar fashion. A signal is added to the noise data at different signal-to-noise ratios in order to determine the minimum SNR necessary to obtain 50% of the signal + noise detector outputs exceeding the detection threshold. The signal used here is an artificial Gaussian signal generated with the polar method of the IMSL statistical package. The measure of SNR considered here is

$$SNR = 10 \log_{10}(\sigma_s^2/\sigma_n^2)$$

where

σ_s^2 = sample variance of the signal

σ_n^2 = sample variance of the noise.

IV. APPLICATION TO OCEAN ACOUSTIC NOISE DATA

In this section, the techniques described above are used to evaluate the performance of the energy detector in the ocean environment. First, a brief description of the data is given. Estimates of the relationship between false alarm rate and detection threshold are displayed and the corresponding MDS estimates are provided for an artificial Gaussian signal.

Ambient noise data from five different noise environments are used in the simulations. A detailed statistical analysis of these data is provided in Ref. 9. A brief description of the data is given below.

1. *Quiet Gaussian Data* - Recorded during a quiet period in the deep Pacific Ocean; passes univariate tests for normality.

2. *Noisy Gaussian Data* - Recorded in the Indian Ocean; no single surface ship dominates the noise field; fails univariate tests for normality due to marginally high kurtosis.

3. *Merchant Vessel Data* - Recorded in the same area as (2) during a time period when a merchant vessel is in close proximity to the recording system. Fails univariate tests for normality because of low kurtosis.

4. *Seismic Data* - Recorded in the Gulf of Mexico during seismic exploration activities; fails tests for normality with high skew and kurtosis; impulsive character of the data is readily apparant in time series.

5. *Antarctic Data* - Recorded in the early spring in McMurdo Sound, Antarctica (Ref. 10); characterized by heavy biological activity; time series reveals strong impulsive components as the data is highly nonstationary and non- Gaussian.

Because the data were collected with different recording systems having different bandwidths, the data are digitized at different sampling rates varying from 892 Hz for the seismic data to 2400 Hz for the Antarctic data. The dependence of the results to follow on the bandwidth of the data is removed by a randomization procedure described below. As a result, no attempt was made to convert the data to the same bandwidth. Prior to input to the energy detector, each data set is normalized to have zero mean and unit variance, where the mean and variance are averaged over the entire data set. Hence, the evaluation is performed on an equal variance basis.

When the data are sampled at the Nyquist rate, sample autocorrelation functions reveal significant correlations at time delays up to 100 samples and longer. The "runs up and down" statistical test for randomness (Ref. 11, 12) was also applied to the data indicating that a randomness hypothesis can be rejected with a high degree of confidence (Ref. 9). Since the energy detector is designed for independent samples, performance measurements may be significantly biased by correlated samples. In this study, the main interest lies in quantifying performance degradation of the energy detector attributable to non- Gaussian noise statistics; hence, uncorrelated noise samples are desired.

In order to obtain uncorrelated samples, the data is randomized in the following way. First a realization consisting of 200,000 samples at the Nyquist rate is segmented into 200 blocks of length 1000. These blocks are then randomly permuted using a permutation vector generated by the IMSL statistical package. The samples from the resulting data set are then shuffled to form 64 groups of data containing 3125 samples in each group. The mth group consists of every 64th sample starting with sample number m from the block permuted data. The 64 groups are then regrouped (but not reordered) into 40 blocks of length 5000. The data within each of these 40 blocks is randomly permuted using different permutation vectors from the IMSL package. After randomization of the data, the corresponding sample autocorrelation functions show no significant correlations. Furthermore, results of the runs test indicate that a randomness hypothesis may be accepted for each of the randomized data sets (see Ref. 9 for details).

In order to estimate the probability of false alarm (PFA) for values of interest (10^{-3} and smaller), a large number of trials are required to obtain meaningful estimates. This requirement precludes the use of long integration times because of the enormous amount of data required. In this study, therefore, a short integration of 10 samples is used, corresponding to a time period of about 10 msec. This choice results in 20,000 detector outputs for each data set. With 20,000 trials, the standard deviation of the PFA estimate is about 7.04×10^{-4} at a PFA of 10^{-2} and about 2.23×10^{-4} at a PFA of 10^{-3} (Ref. 8). Simulations with integrations of 20, 40, and 80 samples were also performed but are not included here for the sake of brevity. The results that follow show very little difference from the results for the longer integration times.

Figure 1 displays PFA estimates (on logarithmic scale) as a function of detection threshold for the five data sets. The theoretical curve for i.i.d. Gaussian noise, derived from the chi-squared distribution with 10 degrees of freedom, is also shown for comparison. The estimates for the quiet Gaussian and noisy Gaussian data are seen to give the best agreement with theory. The required threshold settings for the merchant dominated data are significantly lower than theory. In contrast, the estimates for the seismic and the Antarctic data are considerably greater than theory. At a false alarm probability of 10^{-3}, the estimated threshold settings for the seismic data are about 50% larger than theory. For the Antarctic data, the threshold estimates are more than twice the theoretical value for white Gaussian noise.

Kurtosis estimates for the noise data provide insight into the behavior of Figure 1. The estimate of kurtosis used here is give by

$$\beta_2 = m_4/(m_2)^2$$

where m_k denotes the kth sample moment about the mean. Kurtosis can be used as a measure of Gaussianity giving an indication of the size of the tails of a density function. In general, a larger kurtosis is associated with larger tails. This means that a larger proportion of outliers (large samples) can be expected in data with a large kurtosis than in data with the same variance and a small kurtosis. These outliers inflate the variance of the energy detector output leading to higher threshold settings. They also tend to inflate the skew and kurtosis of the energy detector output.

Table I contains values of the sample kurtosis averaged over 200 data segments of length 1000 for the five noise environments. Table II contains sample variance, skew, and kurtosis values for the energy detector outputs. These values are based on an average over 20 blocks of detector outputs of length 1000. Note the correspondence between the values in the tables and the curves of Figure 1.

TABLE I. Kurtosis measurements for noise data.

DATA SET	KURTOSIS
THEORY	3.00
QUIET GAUSSIAN	2.97
NOISY GAUSSIAN	3.14
MERCHANT	2.29
SEISMIC	5.69
ANTARCTIC	6.02

TABLE II. Sample moments of energy detector output.

DATA SET	VARIANCE	SKEW	KURTOSIS
THEORY	20.0	0.894	4.20
QUIET GAUSSIAN	19.9	0.849	3.96
NOISY GAUSSIAN	20.9	0.954	4.46
MERCHANT	13.0	0.498	3.26
SEISMIC	41.9	1.50	6.76
ANTARCTIC	40.4	3.06	26.5

In Figure 2, estimates of the MDS for an artificial Gaussian signal are displayed for the five noise environments. The values shown are in decibels (dB) relative to the theoretical value for i.i.d. Gaussian noise. An integration of 10 samples is used with threshold settings corresponding to estimated false alarm probabilities of 10^{-1}, 10^{-2}, and 10^{-3}. Each MDS estimate is based on 20,000 signal plus noise detector outputs.

Since the MDS estimates are unbiased, the variance of the estimate provides a measure of the quality of the estimate. An estimate of the variance of the MDS estimate is constructed in the following way. Each noise data set is segmented into 40 blocks containing 5000 samples. An MDS estimate is computed for each block by determining the SNR required to obtain 250 out of 500 signal plus noise detector outputs exceeding the threshold estimate for the PFA of interest. This procedure results in 40 block MDS estimates for which a sample mean and variance are computed. Assuming that the block MDS estimates are independent, the variance of the sample mean is simply the sample variance divided by the number of blocks. The variance of the sample mean should be roughly the same as the variance of the MDS estimate for the entire collection of 20,000 detector outputs. Using this procedure, the largest variance of the overall MDS estimate was about 2.8×10^{-3}, corresponding to a value of 0.16 dB for three standard deviations. Allowing for the variance of the threshold estimate, the MDS estimates are approximately correct to within 0.5 dB for a PFA of 10^{-3} or larger.

Comparison of Figures 1 and 2 reveals the influence of the threshold setting on the ability to detect a signal. The high threshold settings for the seismic and the Antarctic data translate into a loss of 2-4 dB in terms of the MDS at a false alarm probability of 10^{-3}. Note how this loss increases with decreasing probability of false alarm following the curves of Figure 1. In contrast, the MDS estimates for the merchant dominated data are 1-2 dB less than theory. As expected, the MDS estimates for the quiet Gaussian and noisy Gaussian data agree to within 0.25 dB of theory. The MDS estimates should be considered carefully since the noise data are normalized to have equal variance. For instance, the noise levels differed by about 6-10 dB between the merchant dominated data and the noisy Gaussian data collected from the same location. In addition, the quiet Gaussian data was collected from an area of anomalously low ambient noise levels and probably represents one of the most favorable environments for detections to occur.

The above results are dependent on the randomization procedure in a couple of ways. First of all, randomization tends to destroy the temporal correlations in the data. The threshold estimates can be heavily biased by dependent samples simply because the sample variance is affected by dependent samples. The graphs of Figs. 1 and 2 show how the threshold setting strongly impacts the detectability of a signal. Secondly, and perhaps of more importance, the randomization of the data tends to spread out the effects of nonstationarities in data such as the seismic and the Antarctic data. These nonstationarities can lead to a much more serious degradation of detection performance than shown here.

V. CONCLUSION

The performance of the energy detector has been analyzed using recorded ambient noise data from several ocean acoustic environments. Simulations indicate that degraded detector performance can be expected in some non-Gaussian and nonstationary noise fields. This performance loss is linked to the size of the tails of the underlying noise density. In impulsive noise environments, this degradation can be attributed to the heavy tailed densities associated with these environments. This performance loss may be reduced to some extent through the use of robust detection algorithms.

ACKNOWLEDGEMENT

This work was supported by the Statistics and Probability Program of the Office of Naval Research. The authors appreciate the helpful discussions relative to the work with Dr. Gary Wilson and Dr. Patrick Brockett.

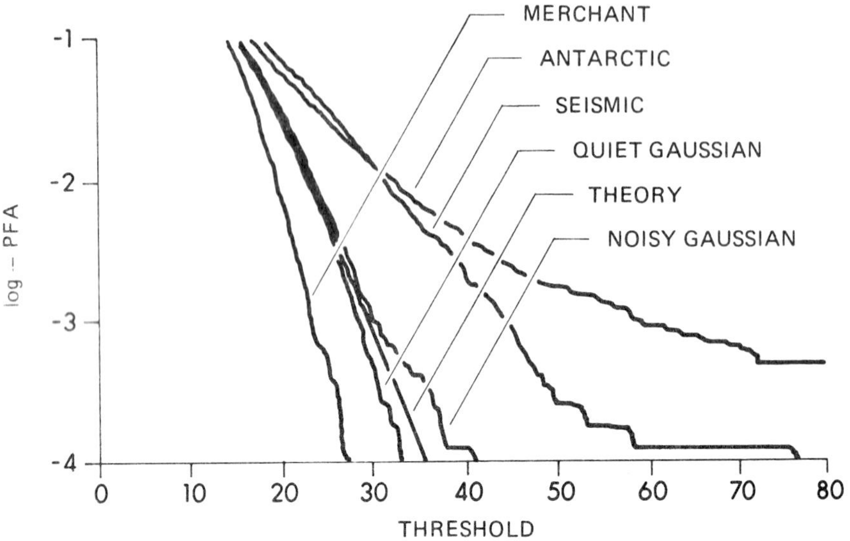

Figure 1 False Alarm Probability Estimates as a Function of Detection Threshold

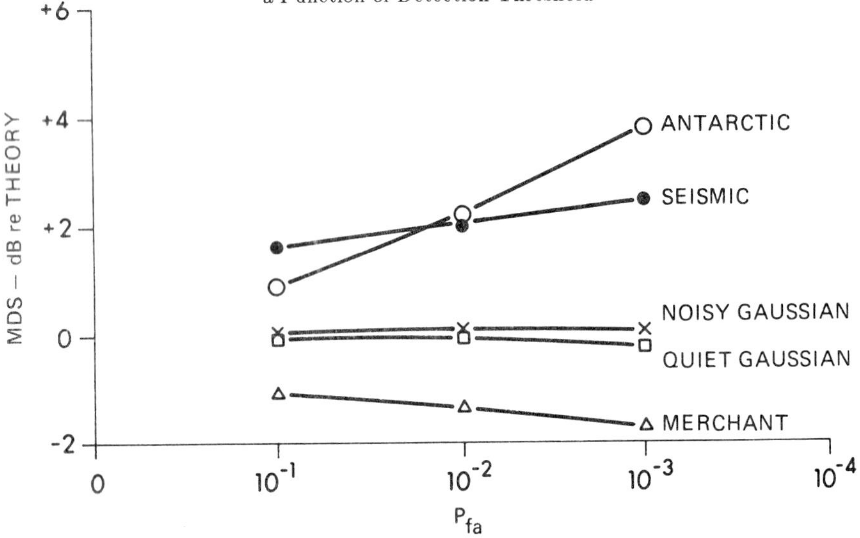

Figure 2 Minimum Detectable Signal-to-Noise Ratio for White Gaussian Signal in dB Relative to the Theoretical Value for White Gaussian Noise

REFERENCES

1. C.W. Horton, Sr., Signal Processing of Underwater Acoustic Waves, U.S. Government Printing Office, Washington, DC, (1969).

2. R.J. Urick, Principles of Underwater Sound, 2nd ed., McGraw-Hill, New York, (1967).

3. A.R. Milne and J.H. Ganton, "Ambient Noise Under Arctic-Sea Ice," *J. Acoust. Soc. Am.,* Vol. 36, p. 855-863, (1964).

4. R. Dwyer, "FRAM II Single Channel Ambient Noise Statistics," *NUSC Technical Document 6583,* (1981).

5. F.W. Machell and C.S. Penrod, "Probability Density Functions of Ocean Acoustic Noise Processes," in *Statistical Signal Processing,* edited by E.J. Wegman and J.G. Smith, Marcel Dekker, New York (1984).

6. J.G. Veitch and A.R. Wilks, "A Characterization of Arctic Undersea Noise," *Dept. of Statistics Technical Report,* No. 12, Princeton Univ., (1983).

7. J.M. Hammersley and D.C. Handscomb, Monte Carlo Methods, Metheun, London (1964).

8. G.W. Lank, "Theoretical Aspects of Importance Sampling Applied to False Alarms," *IEEE Trans. Information Theory,* Vol. IT-29, 73-82, (1983).

9. F.W. Machell, C.S. Penrod, and G.E. Ellis, "Statistical Characteristics of Ocean Acoustic Noise Processes," to appear in *ONR publication on Non-Gaussian Signal Processing.*

10. A.C. Kibblewhite and D.A. Jones, "Ambient Noise Under Antarctic Sea Ice," *J. Acoust. Soc. Am.,* Vol. 59, p. 790-798, (1976).

11. J.V. Bradley, Distribution-free Statistical Tests, Prentice, Englewood Cliffs, NJ (1968).

12. G.R. Wilson, "A Statistical Analysis of Surface Reverberation," *J. Acoust. Soc. Am.,* Vol. 74, p. 249-255, (1983).